Sustainable Development Goals Series

The **Sustainable Development Goals Series** is Springer Nature's inaugural cross-imprint book series that addresses and supports the United Nations' seventeen Sustainable Development Goals. The series fosters comprehensive research focused on these global targets and endeavours to address some of society's greatest grand challenges. The SDGs are inherently multidisciplinary, and they bring people working across different fields together and working towards a common goal. In this spirit, the Sustainable Development Goals series is the first at Springer Nature to publish books under both the Springer and Palgrave Macmillan imprints, bringing the strengths of our imprints together.

The Sustainable Development Goals Series is organized into eighteen subseries: one subseries based around each of the seventeen respective Sustainable Development Goals, and an eighteenth subseries, "Connecting the Goals", which serves as a home for volumes addressing multiple goals or studying the SDGs as a whole. Each subseries is guided by an expert Subseries Advisor with years or decades of experience studying and addressing core components of their respective Goal.

The SDG Series has a remit as broad as the SDGs themselves, and contributions are welcome from scientists, academics, policymakers, and researchers working in fields related to any of the seventeen goals. If you are interested in contributing a monograph or curated volume to the series, please contact the Publishers: Zachary Romano [Springer; zachary.romano@springer.com] and Rachael Ballard [Palgrave Macmillan; rachael.ballard@palgrave.com].

Xinbin Hou

Space Solar Power

An Illustrated Introduction

Springer

Xinbin Hou
Qian Xuesen Laboratory of Space Technology
China Academy of Space Technology
Beijing, China

ISSN 2523-3084 ISSN 2523-3092 (electronic)
Sustainable Development Goals Series
ISBN 978-981-97-3579-2 ISBN 978-981-97-3580-8 (eBook)
https://doi.org/10.1007/978-981-97-3580-8

The original submitted manuscript has been translated into English. The translation was done using artificial intelligence. A subsequent revision was performed by the author(s) to further refine the work and to ensure that the translation is appropriate concerning content and scientific correctness. It may, however, read stylistically different from a conventional translation.

Jointly published with Chemical Industry Press

The print edition is not for sale in China (Mainland). Customers from China (Mainland) please order the print book from: Chemical Industry Press.

This Springer imprint is published by the registered company Springer Nature Singapore Pte Ltd.
The registered company address is: 152 Beach Road, #21-01/04 Gateway East, Singapore 189721, Singapore

This book is dedicated to my parents, my wife, and my daughter, for your selfless support and love. I love you forever.

Foreword

The energy crisis and environmental pollution are major issues that affect global sustainable development and the creation of a community for the shared future of mankind. Achieving net zero emissions as soon as possible has become a global consensus. In September 2020, China officially announced at the 75th United Nations General Assembly that it strives to peak carbon dioxide emissions by 2030 and aims to achieve carbon neutrality by 2060. To achieve the "dual carbon" strategic goals, the world's demand for renewable energy will continue to grow. However, the extreme instability of traditional renewable energy technologies poses a significant challenge to the existing clean energy system in achieving net zero emission goals. It is urgent to develop new large-scale clean energy sources that can serve as a baseload power supply.

Space Solar Power (SSP) can provide massive, continuous, stable, clean, and green electricity, making it a potential key player in the world's clean energy landscape. The SSP system is expected to play a major role in aerospace equipment systems. With advancements in reusable launch technology and the ability to develop large-scale spacecraft, the development of SSP is becoming increasingly significant from both technical and economic perspectives. The United States, Japan, Korea, and some European countries have regarded SSP as an important policy option to meet this challenge. The next 10–20 years will be the important period for rapidly technological breakthroughs, on-orbit demonstration, and application. SSP is facing new development opportunities.

In 1985, Academician Xiji Wang proposed the idea of space energy utilization. In 1992, several Chinese researchers participated in the SSP design activity organized by International Space University (ISU). In 2006, under the promotion of experts such as Academicians Fenggan Zhuang, Xiji Wang, and Guirong Min, China began to initiate research on SSP. In 2013, Academicians Shizhong Yang and Baoyan Duan proposed to the state to develop SSP, which received high attention from the government. In 2014, the China National Space Administration, in conjunction with the National Development and Reform Commission, the Ministry of Science and Technology, and some other departments, organized the expert team to carry out studies and proposed suggestions for China's SSP development roadmap.

In recent years, China's research on SSP has been steadily advancing, initially forming a series of system solutions and key technological innovation achievements, and the overall research level have been improved continuously.

The development of SSP involves a diverse range of professions and fields of expertise, including mechanics, electronics, microwave technology, laser technology, thermal physics, control systems, materials, as well as aerospace, energy, electricity, and other related fields. It also touches upon environmental, safety, policy, legal, and other related areas. The SSP system is a major scientific and technological project in the aerospace industry, driving innovation in frontier technologies such as space ultra-high voltage and high-power electric generation, space ultra-large structure on-orbit construction and control, and space long-distance efficient wireless power transmission. Additionally, SSP has the potential for applications beyond commercial power generation, including power supply in remote areas, mobile power supply, disaster relief power supply, deep space power supply, and more, offering a wide range of application value and significant social and economic benefits.

With strong support from leaders and experts, the Committee of Space Solar Power of the Chinese Society of Astronautics (CSSP-CSA) was officially established in March 2021, providing a high-level academic exchange platform for Chinese and international scholars involved in SSP research. Through technical collaboration and industry-university-institute cooperation, the committee aims to cultivate world-class scientific and technological leaders and innovative teams, driving the progress and commercialization of SSP. At the same time, promoting the knowledge of SSP to the public is also an important mission of the Committee. The author of this book, a member of the Committee who has long been engaged in the research in this field, is an internationally renowned expert with his deep understanding and insights of SSP. As a new attempt, this book systematically presents a combination of illustrations and text that vividly describes the development background, current status, conceptual scheme, and main technologies of SSP. It is intended for various reader groups, enabling the public to better understand the development significance, prospects, and core technical issues of SSP. This book is the first semi-popular science work on SSP and will play a positive role in popularizing, promoting, and disseminating knowledge about SSP to the public and cultivating experts for basic research and technical R&D.

We look forward to the support of all sectors of society for the development of SSP and hope that more experts and young scholars can devote themselves to the R&D and innovation work of SSP. We also welcome those who are interested to join the CSSP, to jointly create a social and scientific and technological ecosystem for the development of SSP, and promote the early realization of SSP to benefit mankind.

CSSP-CSA Ming Li
Beijing, China

Preface

Energy is the main basis for human society to survive and develop. The global energy consumption is increasing quickly. The global warming trend brought by the consumption of traditional fossil energy is obvious, and extreme weather is frequent, which greatly affects human survival. The energy crisis has become one of the core issues restricting the sustainable development of the world's society and economy. Developing clean energy to gradually replace traditional fossil energy and achieve net zero emissions as soon as possible has become a global consensus. In September 2020, China proposed to strive to achieve carbon peak by 2030 and carbon neutrality by 2060. Due to the extreme instability of renewable energy technology, based on the existing clean energy system to meet human demand for reliable energy and achieve the global carbon neutrality goal faces huge challenges. It is urgent to develop new large-scale clean energy technologies that can serve as baseload power sources.

Space solar power (SSP) is a space engineering project that collects and converts solar energy on a large scale in space and transmits electric energy to the ground grid wirelessly. Its unique advantage is that it is not affected by day and night and weather changes, providing sustainable clean energy for mankind. It is a grand plan for mankind to use space resources to solve future energy crises and environmental problems. In recent years, aerospace technology has developed rapidly, especially the reusable launch technology has greatly reduced the cost of space transportation, and the scaled spacecraft development model has greatly reduced the cost of spacecraft manufacturing. The development of SSP is gradually becoming of great practical significance from a technical and economic perspective. The value of SSP has been being re-emphasized by the United States, Japan, ESA (European Space Agency), South Korea, the United Kingdom, Australia, and other countries and related international organizations and private companies. China has also carried out long-term research and has become a core force in promoting the development of SSP. SSP is expected to become an important part of the carbon neutrality goal and the clean energy system for dealing with global climate change in the future.

This book is based on the author's research experience in the field of SSP, referring to relevant international references, and using a combination of pictures and texts, starting from the background of global climate change and carbon neutrality goals. It introduces the development status of international SSP comprehensively, analyzes and compares internationally various concepts and schemes, and elaborates on the space high-power solar power generation, wireless power transmission, the transportation and on-orbit construction modes of SSP, and systematically introduces the Multi-Rotary joint Solar Power Satellite (MR-SPS) concept proposed by China. The book is divided into seven chapters. Chapter 1 mainly introduces the development background and international development status of SSP; Chap. 2 introduces the operating orbit, space environmental characteristics, system composition, etc. of the SSP system; Chap. 3 describes and classifies the typical SPS types proposed internationally; Chap. 4 introduces the main technologies involved in the space high-power solar power generation system; Chap. 5 introduces the unique wireless power transmission system of SSP; Chap. 6 focuses on the possible transportation modes and on-orbit assembly requirements of the SPS; Chap. 7 introduces the system composition and schemes of typical SSP system represented by MR-SPS.

During the writing of this book, the research results of global researchers were referred to. I would like to express my sincere thanks to my peer around the world and special thanks to the research experts for their long-term support and help!

Due to the limited knowledge and professional scope of the author, there are inevitably shortcomings in the book, and I sincerely hope readers to criticize and correct.

Beijing, China Xinbin Hou

Brief Introduction

Space solar power (SSP) system, a major type of space-based power-generating equipment, is expected to be an important infrastructure providing massive, continuous, and stable green electricity by utilizing solar energy in space. With the fast advances in reusable launch vehicles and heavy-load spacecraft building, developing SSP has become a great practical significance from the technical and economic perspectives. It will be a flexible energy source to supply power to most areas in the world, especially to remote areas where traditional energy grids cannot reach. This book, with the background of global climate change and carbon neutrality goals, combining illustrations and text, mainly introduces the development history, global status, and typical concepts of SSP. It also elaborates key technologies, such as that of high-power solar power generation in space, wireless power transmission, and the transportation and on-orbit construction of solar power satellites (SPS). Finally, the system scheme is introduced based on the Multi-Rotary joint Solar Power Satellite (MR-SPS) proposed by China, which has been one of the typical concepts worldwide.

The content of this book is comprehensive and richly illustrated, which makes it easy for readers to develop a quick and systematic understanding of the history and current status of SSP, its technical characteristics, system composition, and related technologies. The book is suitable for college students specializing in aerospace, energy, electric power, mechanics, electronics, etc., and can also be used as a reference for researchers involved in the study of SSP.

Contents

Background and Development Status of Space Solar Power

1

1.1 Overview

Energy is the main material basis for human society to survive and develop. With the rapid increase in energy consumption, energy shortage and global climate and environmental problems caused by the consumption of traditional fossil energy have become extremely important issues restricting the sustainable society and economy development of the world. Developing clean energy, exploiting renewable energy, and gradually replacing traditional fossil energy have become the global consensus. On November 4, 2016, the "Paris Agreement" officially came into effect, marking an important step for mankind towards the goal of sustainable development. Whether we can achieve "net zero" as soon as possible has become the key to global climate change control. China is the fastest developing country in renewable energy in the world, with solar power and wind power installed capacity ranking first in the world. In September 2020, China proposed to strive to achieve carbon peak by 2030 and carbon neutrality by 2060, reflecting China's great responsibility for building a community of shared future for mankind.

In 2022, nearly all the world experienced rare high temperatures and droughts in history. The obvious global warming trend and frequently extreme weather greatly affect human survival. Europe experienced extreme high temperatures and the most severe drought in 500 years. London saw a record high temperature of 40 °C. The high temperatures and drought caused a significant drop in Europe's hydroelectric power generation. At the same time, Europe is facing a serious energy crisis with soaring electricity prices. Countries represented by Germany are restarting coal-fired power plants that were prepared to be shut down and extending the lifespan of nuclear power plants. China also experienced extreme high temperatures and drought in 2022. Chongqing saw a record 22 consecutive days of extreme high temperatures above 40 °C, with the highest temperature reaching a record 45 °C. Meanwhile, the middle and lower reaches of the Yangtze River experienced a rarely severe drought, with the water area of Poyang Lake shrinking by nearly 90%, revealing a unique "Tree of the Land" landscape (Fig. 1.1). Sichuan Province, China's largest hydroelectric power province and the largest hydroelectric power output province, suffered a rare power shortage in 2022, which greatly affected industrial and agricultural production and people's normal life. Currently, in the context of the annual increase in global energy demand, due to the extreme instability of renewable energy technology, before the establishment of large-scale energy storage facilities, meeting human demand for clean and reliable energy and achieving global carbon neutrality goals based on the existing clean energy technologies face huge challenges. It is urgent to develop new, large-scale clean energy

X. Hou, *Space Solar Power*, Sustainable Development Goals Series,
https://doi.org/10.1007/978-981-97-3580-8_1

Fig. 1.1 Tree of the land (Poyang Lake)

technologies that can serve as baseload power sources.

Space solar power (SSP), as the major aerospace equipment system serving the energy sector, will become important infrastructure for human utilization of space solar energy. SSP is expected to provide humanity with abundant, continuous, and stable clean green electricity, which is of great value in solving the world's energy and environmental problems. It will also be an important infrastructure for human to explore the universe and exploit space material resources. SSP has attracted the attention of major space nations internationally, and research in this field has been ongoing for over 50 years worldwide. In 2018, the 50th anniversary conference for space solar power was held in the United States (Fig. 1.2). In recent years, with the rapid progress of aerospace technology, especially the significant reduction in space transportation costs due to reusable launch technology, and the massive spacecraft development model greatly reducing the manufacturing cost of spacecraft, SSP is gradually becoming of great practical significance from the technical and economic perspective. SSP is once again receiving widespread attention from the United States, Japan, ESA (European Space Agency), South Korea, the United Kingdom, Australia, and other countries, as well as relevant international organizations and private companies. The United States is launching the new round of R&D projects. Japan has listed SSP as a key development direction in its Space Basic Plan. ESA has included SSP in ESA Agenda 2025. The UK has included the development of SSP in its National Space Strategy and Net Zero Innovation Portfolio. South Korea has launched related research project with the support of the parliament. China has also conducted long-term research in the field of SSP and has become a core role in promoting the development of SSP. SSP is meeting great development opportunities and is expected to become an important part of the clean energy system for addressing global climate change and carbon neutrality in the future. According to a research report by Citibank, the market size of SSP is expected to reach $23 billion by 2040.

Fig. 1.2 Poster of the 50th anniversary of SSP conference

1.2 Global Climate Change and Carbon Neutrality Goals

1.2.1 Greenhouse Effect and Its Impact

As a planet suitable for human habitation, the moderate temperature of Earth is one of the most important factors, which is determined by the relevant energy balances. The main source of heat on the earth comes from the sun, a star that stably undergoes nuclear fusion, with a temperature of about 6000 K. The corresponding blackbody irradiation spectrum (AM0) of the sun is shown in Fig. 1.3, mainly including three areas: the ultraviolet region, corresponding to a wavelength of 300–400 nm, accounting for 5% of the energy; the visible light region, corresponding to a wavelength of 400–700 nm, accounting for 43% of the energy; the infrared region, corresponding to a

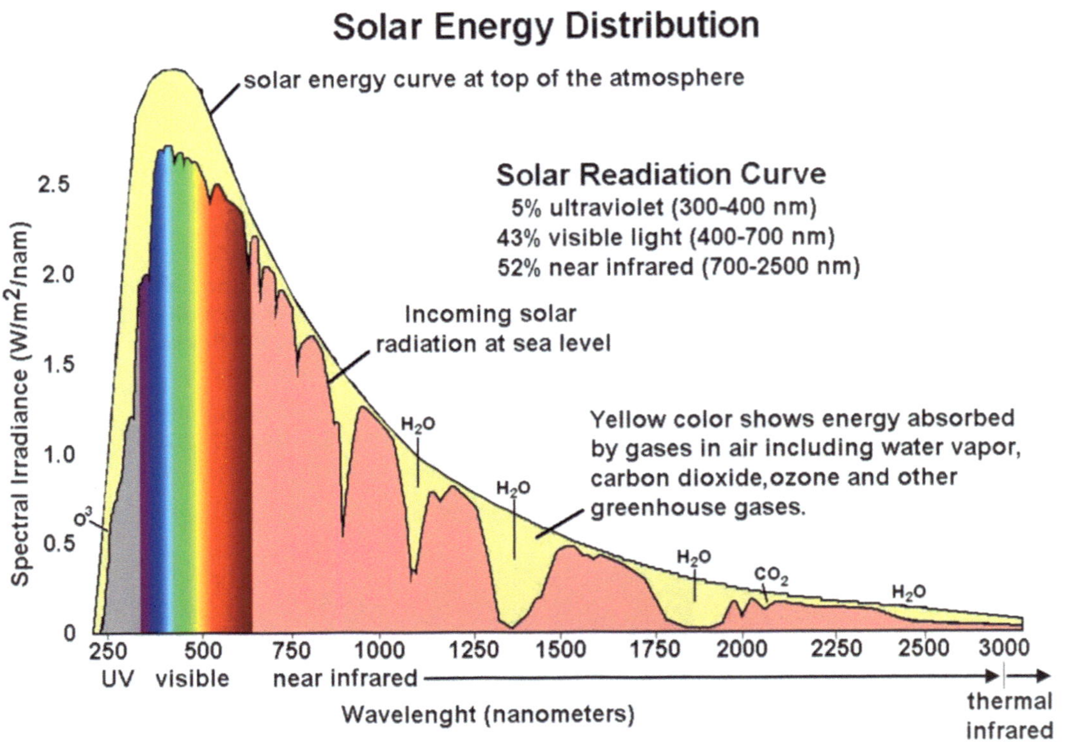

Fig. 1.3 Diagram of solar spectrum distribution in space

wavelength of 700–2500 nm, accounting for 52% of the energy. The solar radiation intensity above the earth's atmosphere is approximately 1360 W/m^2 after traveling a distance of about 150 million km. Due to the absorption of energy in specific spectral bands by water molecules, carbon dioxide, ozone, etc. in the earth's atmosphere, the intensity of solar radiation will also decrease accordingly. The actual solar spectrum (AM1.5) on the ground is shown in Fig. 1.4. At the same time, the earth mainly dissipates heat to outer space through infrared thermal radiation. If there is no earth's atmosphere, the total amount of solar light absorbed by the earth's surface and the heat radiation corresponding to the earth's temperature to space will maintain a balance, and the corresponding ground temperature should be about −18 °C. However, the actual average temperature of the earth is about 15 °C, and the main reason for which is the greenhouse effect produced by the earth's atmosphere. Sunlight can pass through the atmosphere to illuminate the ground well, but the ground infrared thermal radiation is just in the spectral band where water molecules, carbon dioxide, methane, etc. absorb most severely. Therefore, the ground infrared thermal radiation cannot directly pass through the atmosphere to exchange heat with space, which raises the ground temperature from the −18 °C without atmosphere to the current 15 °C. The greenhouse effect of the atmosphere is an important factor for the earth to become a habitable planet.

The energy transfer process of the earth is shown in Figs. 1.5 and 1.6. The incident sunlight intensity of is 340 W/m^2 (global average). Among them, clouds and atmosphere reflect 22.6%, the ground reflects 6.7%, the atmosphere absorbs 22.7%, and the earth actually absorbs 48%, corresponding to 164 W/m^2. The thermal radiation intensity of the ground is 398 W/m^2, of which 53 W/m^2 directly radiates to space through the atmosphere, and the heat dissipated by the ground through conduction, convection, and evaporation

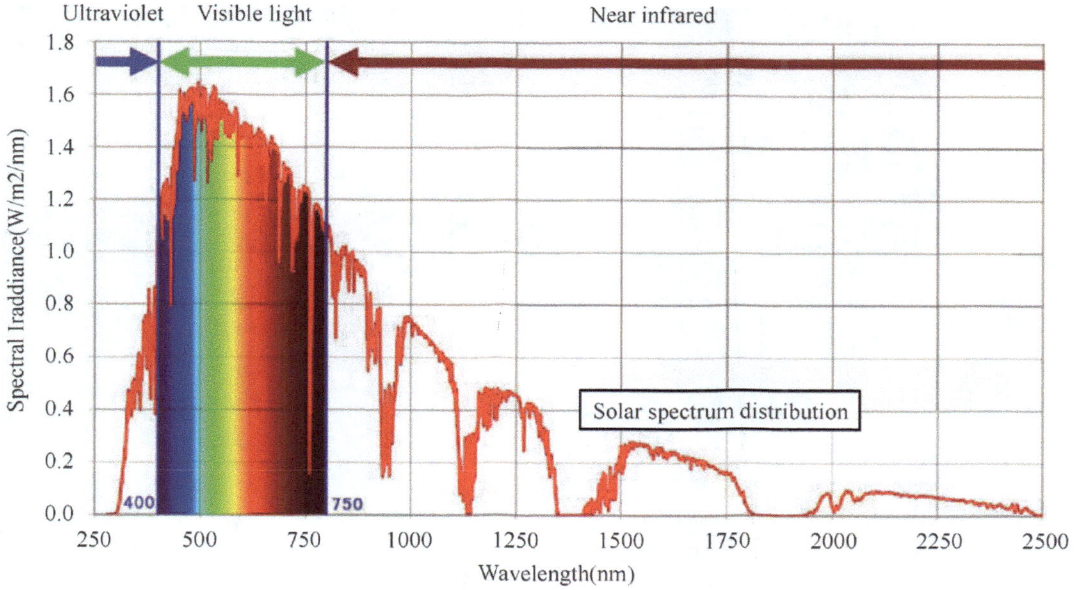

Fig. 1.4 Diagram of solar spectrum distribution on the ground

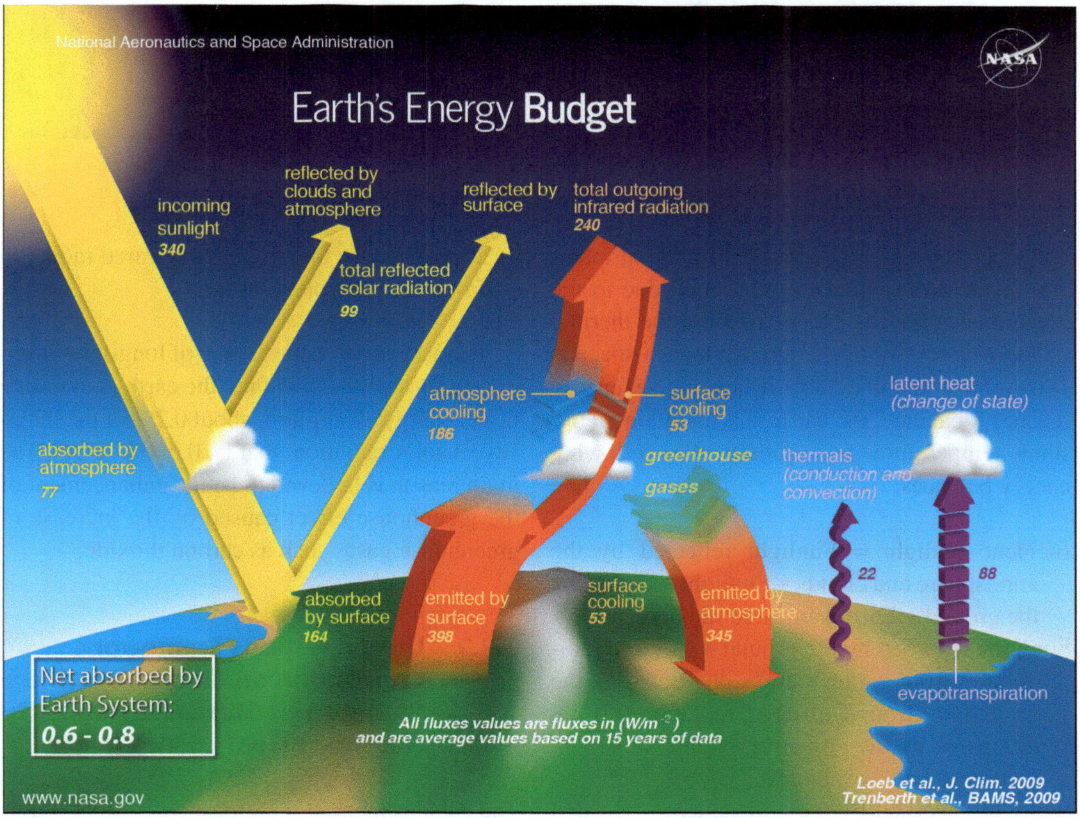

Fig. 1.5 Earth's energy budget (by Power). (Credit: NASA)

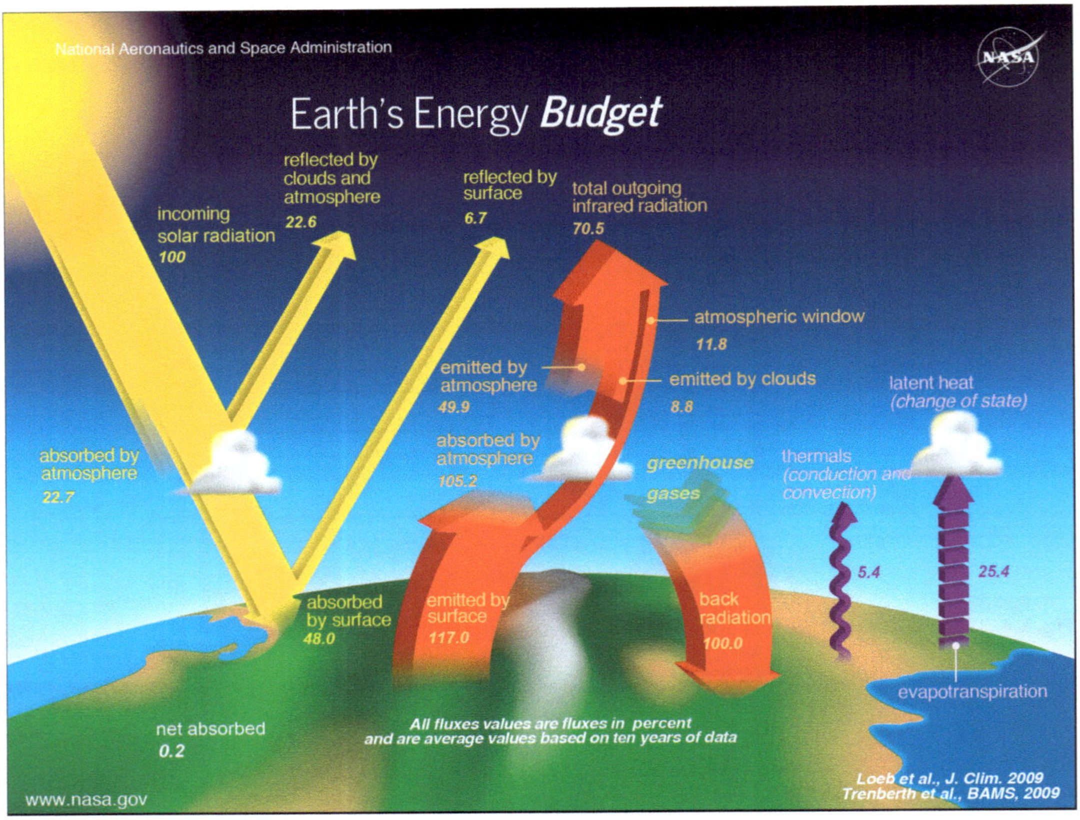

Fig. 1.6 Earth's energy budget (by Proportion). (Credit: NASA)

is about 110 W/m^2. The thermal radiation intensity of the atmosphere to the ground due to the greenhouse effect is 345 W/m^2, and the thermal radiation intensity of the atmosphere to space is 186 W/m^2. Theoretically, if the above energy reaches a balance, the earth will maintain a constant temperature and satisfy the following energy balances:

Incident sunlight = Sunlight reflected by the atmosphere and clouds + Sunlight reflected by the ground + Earth's infrared radiation.

Sunlight absorbed by the ground + Infrared radiation from the atmosphere to the ground = Ground infrared radiation + Heat dissipated by the ground through conduction, convection, and evaporation.

Sunlight absorbed by the atmosphere + Ground infrared radiation + Heat dissipated by the ground through conduction, convection, and evaporation = Infrared radiation from the atmosphere to the ground + Infrared radiation from the atmosphere to space.

But according to the results of long-term monitoring by scientists, in fact, the earth is in a state of net energy absorption, about 0.2% of the intensity of solar radiation, that is, 0.6–0.8 W/m^2. The main reason is believed to be the enhancement of the greenhouse effect caused by the increase of greenhouse gases such as carbon dioxide.

1.2.2 United Nations Framework Convention on Climate Change and Global Carbon Neutrality Goal

The enhancement of the greenhouse effect causes the temperature of the earth's surface to be in a gradually rising state. Global warming will have a significant impact on the earth's environment and ecology and the direct result is a significant increase in the temperature of the North and South Poles, causing glaciers to melt, thereby causing sea level rise. At the same time, it will greatly affect the earth's atmospheric circulation and ocean currents, resulting in abnormal climate changes and a significant increase in extreme weather. Through research, scientists have found that the main reason for global warming is likely to be the rapid industrialization process over the past 100 years since the Industrial Revolution. Energy is the main material basis for human society to survive and develop. The process of industrialization has greatly increased human productivity. But in the meantime, the consumption of energy has grown geometrically. Especially traditional fossil fuels such as coal and oil occupy a large proportion and emit a large amount of carbon dioxide and other greenhouse gases. At the same time, the destruction of primary forests and vegetation also leads to a reduction in plant absorption of carbon dioxide, ultimately causing the concentration of carbon dioxide in the atmosphere to be in a gradually increasing trend. The Intergovernmental Panel on Climate Change (IPCC) Fifth Assessment Report provides the change curves of surface average temperature, sea level change, and atmospheric carbon dioxide concentration in recent years (Fig. 1.7), showing a clear positive correlation between them. According to this trend, the global climate and environment deterioration brought by global warming will pose a huge threat to the sustainable development of societies and economies worldwide.

To address climate change, the United Nations passed the United Nations Framework Convention on Climate Change (UNFCCC, abbreviated as "Convention"), which came into effect on March 21, 1994, has the ultimate goal of maintaining the concentration of atmospheric greenhouse gases at a stable level to prevent the climate system from being threatened. The Convention is the world's first international treaty aimed at comprehensively controlling the emissions of greenhouse gases such as carbon dioxide and addressing the adverse effects of global warming. In 1997, the third Conference of the Parties to the Convention adopted the Kyoto Protocol, which aimed to limit the greenhouse gas emissions of developed countries to curb global warming, and officially came into effect on February 16, 2005. On December 12, 2015, the milestone, Paris Agreement was adopted at the 21st United Nations Climate Change Conference and officially came into effect on November 4, 2016, marking an important step towards the goal of sustainable development. The long-term goal of the Paris Agreement is to limit the increase in global average temperature to within 2 °C compared to the pre-industrial period, and strive to limit the temperature increase to within 1.5 °C, with the early realization of global net zero emissions becoming a global effort (Fig. 1.8 shows the IPCC's predictions for global warming under different scenarios). In September 2020, at the general debate of the 75th United Nations General Assembly, China solemnly announced that it will enhance its nationally determined contributions, adopt more powerful policies and measures, strive to peak carbon dioxide emissions by 2030, and strive to achieve carbon neutrality by 2060. To achieve human emission reduction measures, vigorously developing clean energy, developing renewable energy, and gradually replacing traditional fossil energy have become the global consensus.

1.2.3 Current Status of New Energy Development

In order to achieve the global net zero emission target and control the earth's temperature rise within 1.5 °C, the International Energy Agency (IEA) predicts the future trend of global power supply methods as shown in Fig. 1.9. Currently, fossil fuels power accounts for nearly 60%, nuclear power and hydropower account for about 30%, while solar PV and wind power account for less than 10%. By 2050, the total amount of electricity supply will double, with solar PV and wind power increasing to 65%, nuclear power and hydropower maintaining at 30%, and others accounting for about 5%.

In terms of total energy supply, for the net zero emission scenario, the International Energy

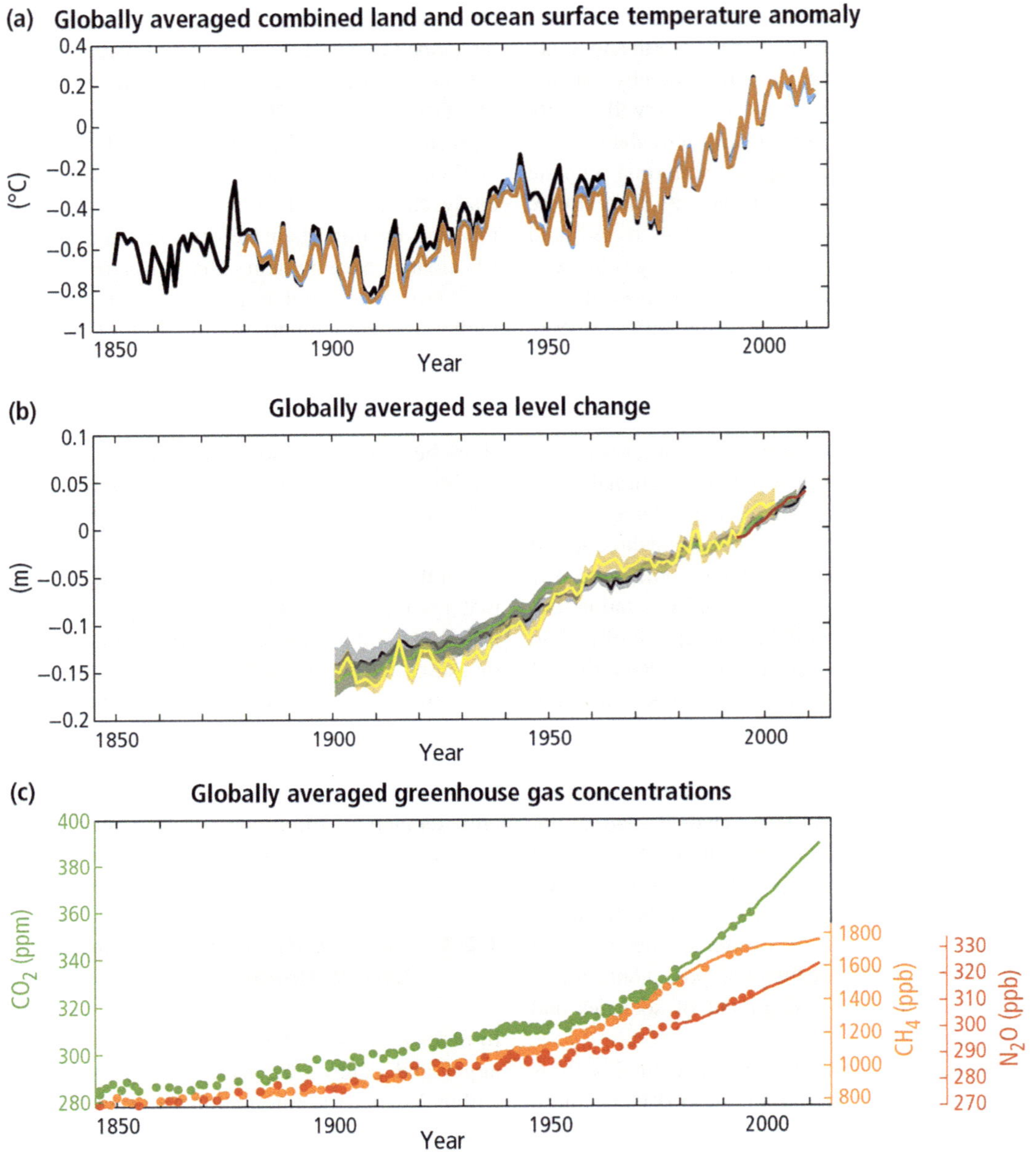

Fig. 1.7 Curve of changes in average surface temperature, sea level changes, and carbon dioxide concentration in the atmosphere. (**a**) Globally averaged combined land and ocean surface temperature anomaly. (**b**) Globally averaged sea level change. (**c**) Globally averaged greenhouse gas concentrations. (Credit: IPCC)

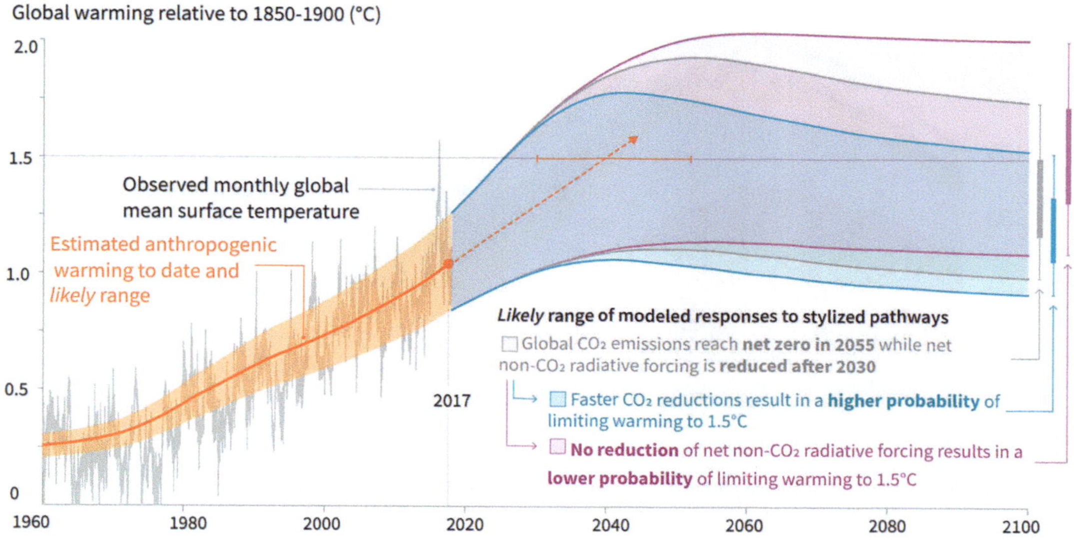

Fig. 1.8 IPCC global warming prediction. (Credit: IPCC)

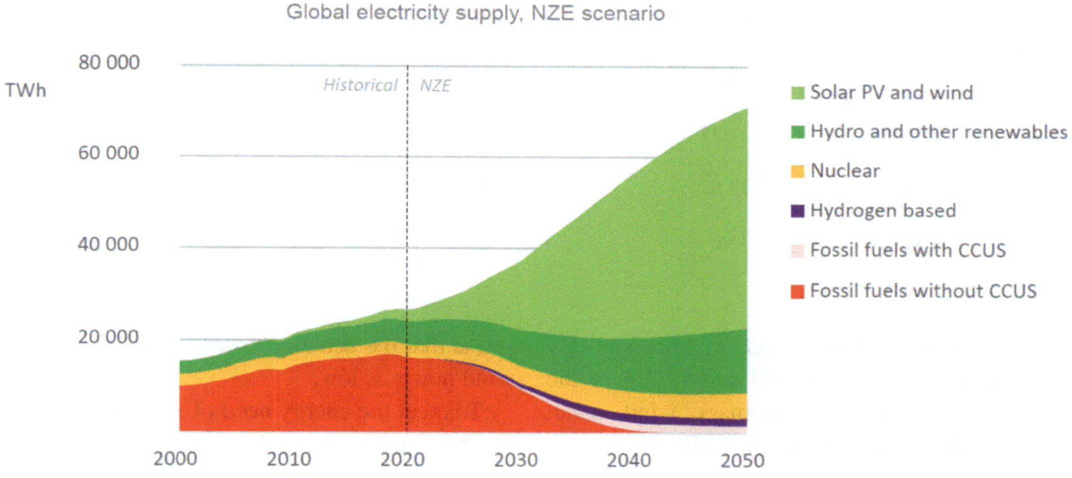

Fig. 1.9 Predicted trend of future global power supply by IEA. (Credit: IEA)

Agency (IEA) predicts future trend as shown in Fig. 1.10. Due to the decrease in energy consumption per unit of production value, the total energy consumption in 2050 will be lower than in 2020, but the energy structure will undergo significant changes. In 2020, oil accounted for 30% of the total energy supply, coal accounted for 26%, natural gas accounted for 23%, and the rest

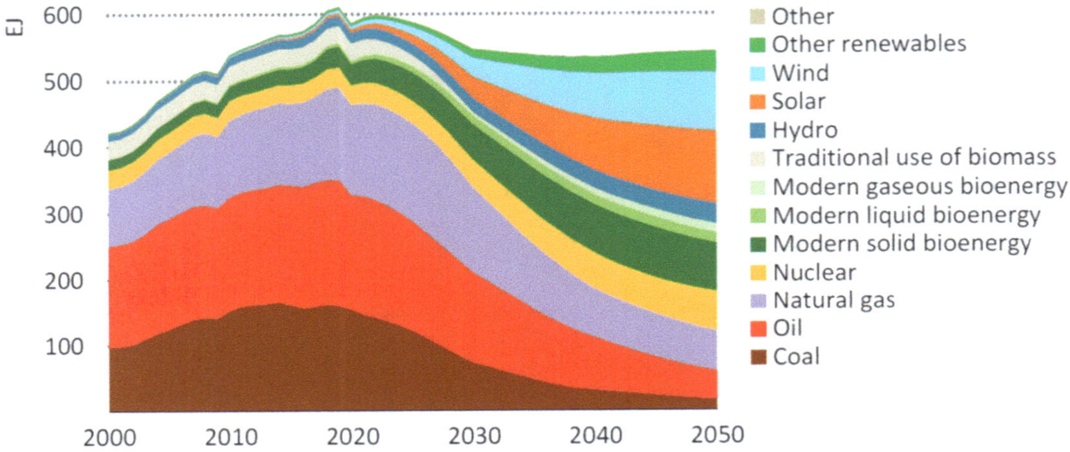

Fig. 1.10 Predicted trend of future global energy consumption by IEA. (Credit: IEA)

Table 1.1 Installed capacity, power generation, and utilization rate of different power generation methods in China (2021)

Power source	Installed capacity/ billion kWs	Proportion of installed capacity/%	Power generation/ billion kWhs	Proportion of power generation/%	Utilization rate/%
Coal power	1.3	54.55	5770.2	71.13	50.67
Hydro power	0.39	16.35	1184.0	14.6	34.65
Nuclear power	0.05326	2.23	407.5	5.02	87.34
Wind power	0.33	13.87	566.7	6.99	19.60
Solar power	0.31	13.0	183.6	2.26	6.76

were nuclear, biomass, and hydropower. By 2050, renewable energy will supply 2/3 of the energy usage, including biomass, wind, solar, hydro, and geothermal energy, while the proportion of fossil fuel usage in the total energy supply will decrease from 80% in 2020 to 20% in 2050.

China is the fastest developing country in renewable energy in the world, with the installed capacity of solar and wind power generation ranking first in the world. Table 1.1, Figs. 1.11 and 1.12 show the installed capacity, power generation, and estimated utilization rate for various power generation methods in China in 2021. It can be seen that the proportion of solar and wind power in total power generation is still very low, in addition to the reason that the total installed capacity is not very high, the most important reason is that the actual utilization rate of solar and wind power is low.

To meet the energy need of sustainable development, humans have developed a large number of clean energy technologies, but the intermittence makes the large-scale development and utilization of clean energy currently very difficult. Although humans have pinned their hopes for the future on vigorously developing solar energy and wind energy, the use of solar energy on the ground cannot avoid the effects of day and night, weather, seasons, and regional latitudes, and wind energy is also greatly affected by weather, seasons, and regions. Relying solely on solar energy and wind energy to replace traditional fossil energy to provide a continuous and stable baseload energy supply for the world still poses a

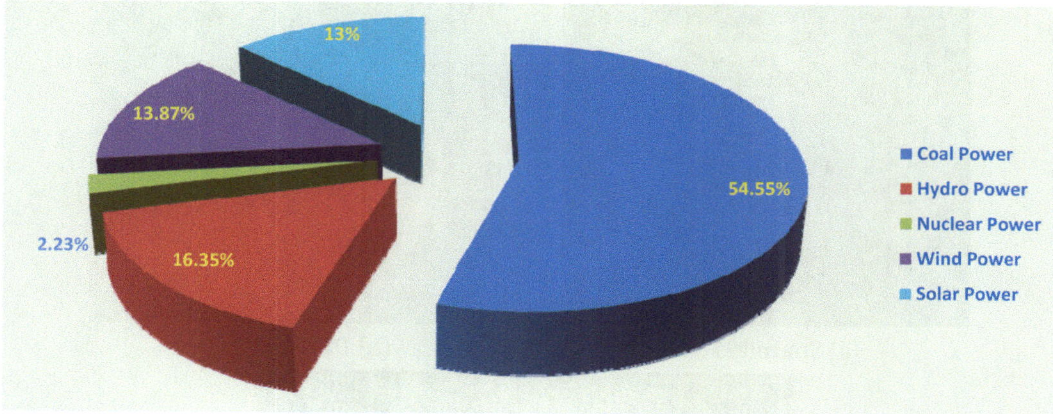

Fig. 1.11 Proportion of installed capacity of different power generation methods in China

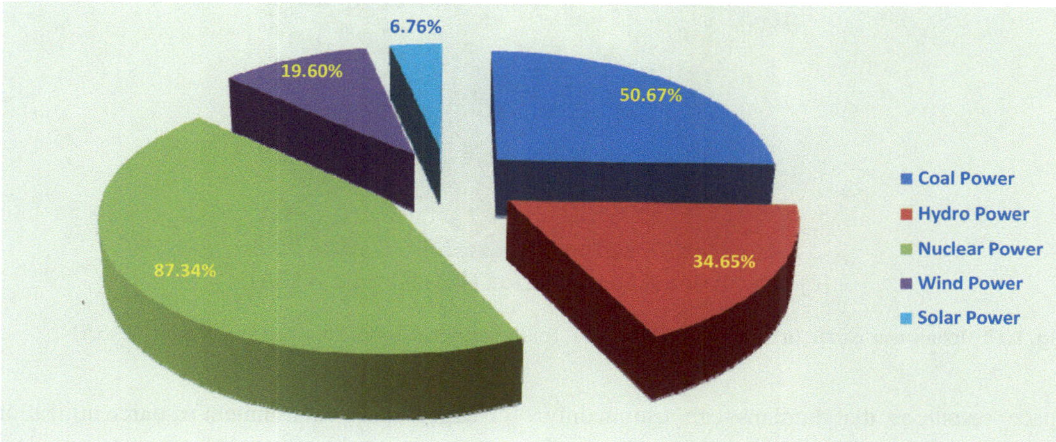

Fig. 1.12 Proportion of power generation by different power generation methods in China

great challenge, which forces people to change their thinking and extend the way to solve the global warming problem from the ground to the outer space.

1.3 Solving Global Warming Issues with Space Technologies

1.3.1 Possible Space Solutions for Global Warming

On October 4, 1957, the Soviet Union successfully launched the world's first artificial satellite, Sputnik-1, marking officially mankind entering into space age and the beginning of human exploration and use of space. The rapid development of aerospace technologies, from space environment detection to the exploration of major planets in the solar system, from space station construction to manned moon landing, from remote sensing, communication to global navigation, has established a strong ground and space infrastructure. Iconic spacecraft are shown in Fig. 1.13.

Whether aerospace technology, as a cutting-edge technology field, can provide strong support for solving global warming problems, needs to be analyzed from the uniqueness of space. The

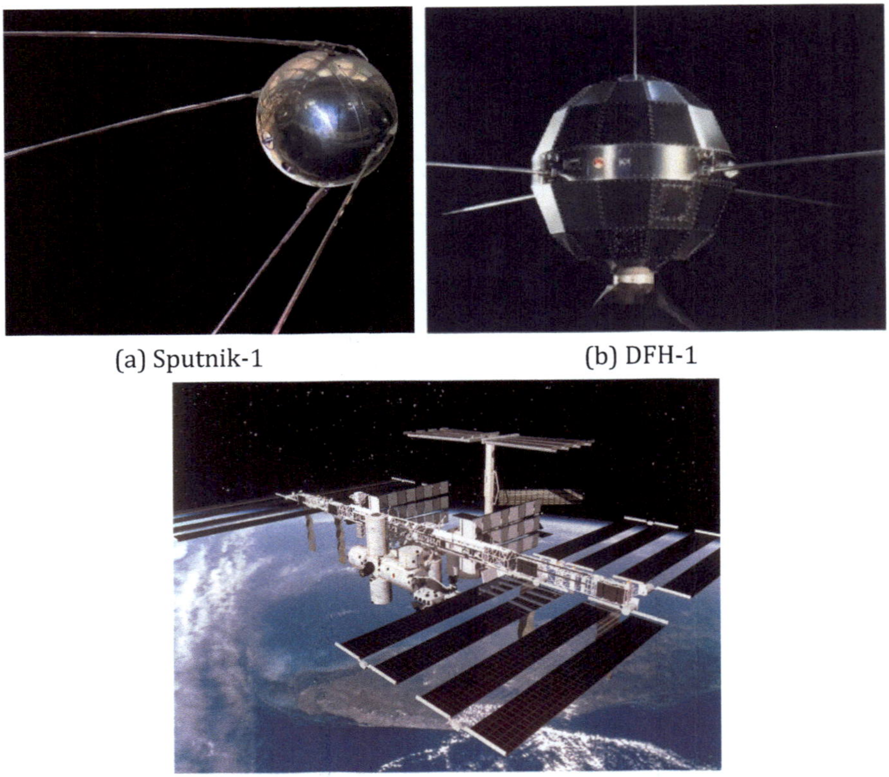

(a) Sputnik-1 (b) DFH-1

(c) International Space Station (ISS) (Credit: NASA)

Fig. 1.13 Iconic spacecraft. (**a**) Sputnik-1, (**b**) DFH-1, (**c**) International Space Station (ISS). (Credit: NASA)

space resources that humans can use mainly include orbit resources, environment resources, material resources, and energy resources.

1. Orbit resources. Orbit resources are the most widely used resources in human space activities, including low Earth orbit, sun-synchronous orbit, geostationary orbit, large elliptical orbit, and interplanetary orbit, etc. Using the height advantages and characteristics of different orbits, they can provide comprehensive information services for the world and has been widely used in communication, navigation, remote sensing, meteorology, scientific detection, etc.
2. Environment resources. Environment resources include the vast space, the special vacuum, microgravity, radiation, low-temperature environment, etc., mainly used for human space science experiments. The International Space Station is currently the largest space environment resource utilization experimental facility.
3. Material resources. Material resources mainly include various materials contained in extraterrestrial bodies such as the moon, planets, and asteroids. With the development of space technology, humans may carry out large-scale space resource utilization in the future, providing support for space exploration, space development, and Earth applications.
4. Energy resources. Energy resources mainly refer to solar energy. At present, almost all spacecraft obtain power through solar cells, greatly increasing the on-orbit lifespan and service capability of the spacecraft.

From the perspective of orbit resource utilization, space technology can provide important information support to solve the global warming problem. Carbon satellites are a type of satellite specifically serving global CO_2 monitoring. They

Fig. 1.14 Schematic diagram of Tansat

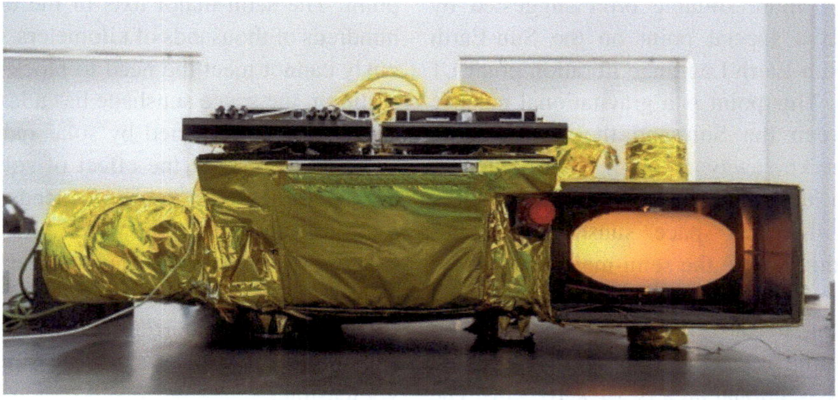

Fig. 1.15 Main payload of Tansat

continuously obtain high-precision CO_2 concentration distribution data on a global scale through passive or active remote sensing, provide important support for global atmospheric environment research and governance, and achieve greenhouse gas emission reduction. On December 22, 2016, China's first carbon satellite (Tansat) was successfully launched (Fig. 1.14), becoming the third greenhouse gas monitoring satellite internationally. Tansat operates in a 700 km sun-synchronous orbit, and the atmospheric CO_2 inversion accuracy is better than 4×10^{-6}. To achieve this indicator, Tansat's main payload is a high-spectral and high spatial resolution CO_2 detector (Fig. 1.15), which achieves high-precision CO_2 detection through high-spectral resolution molecular absorption lines in the visible light and near-infrared spectrum, with its highest spectral resolution reaching 0.04 nm. In April 2018, the first global carbon dioxide distribution map obtained by Tansat was officially released.

From the perspective of environment resources, by reflecting or diverging some sunlight on the path of sunlight propagation in outer space, and blocking some solar radiation energy, we can directly reduce the solar energy absorbed by the Earth, thereby solving the global warming problem. For this, scientists have proposed the concept of space sunshade. According to Fig. 1.6, the net energy absorbed by the Earth is about 0.4% of the solar radiation absorbed by the earth's surface. If we can reduce solar radiation by about 0.4%, we can achieve the energy balance of the Earth. Therefore, theoretically, deploying a super-large space sunshade to directly block sunlight is a possible solution. The first factor to consider is the orbit where the space sunshade is deployed. If the space sunshade is deployed in the earth's orbit, it cannot continuously block sunlight due to its constantly changing position relative to the Sun. Therefore, deploying a huge number of super-large space sunshades orbiting the Earth in space will pose a huge threat to space facilities and normal space activities. A more suitable orbit suggested by scientists is a special point on the Sun-Earth line—the Sun-Earth Lagrange libration point L1 (Fig. 1.16). This point is a gravitational balance point between the Sun and the Earth, about 1.5 million km away from the Earth, and is exactly on the path of sunlight propagation. Assuming that the space sunshade operates within a range where the semi-major axis of the orbit does not exceed the earth's radius around the L1 point, it can effectively block sunlight. The total area corresponding to 0.4% of the earth's cross-sectional area is 5.1×10^5 km^2. The space sunshade can be designed as a super-large thin-film spacecraft (reflecting or scattering sunlight), similar to the concept of a solar sail (Fig. 1.17). Assuming a thin-film spacecraft is 1 km^2, using ultra-thin materials (such as commonly used Kapton film in space), with a thickness of 5 μm, the corresponding area density is about 8 g/m^2. Without considering the mass required for the deployment structure and control auxiliary systems, the mass of a space sunshade is 8 t, corresponding to a total mass of 4.08×10^6 t for 5.1×10^5 km^2, which is thousands of times the total global launch amount per year. The actual payload capacity corresponding to the launch into the Sun-Earth L1 point orbit is much smaller than the payload capacity for launch into low Earth orbit. Operating within a semi-major axis of 6370 km is a significant constraint factor. Generally, to reduce the need for orbit maintenance, spacecraft operating at the L1 point will round an orbit called the halo orbit, which is a non-standard large elliptical orbit around the L1 point. The semi-major axis of the orbit reaches hundreds of thousands of kilometers, which obviously cannot meet the need to block sunlight. In addition, the space sunshade has a large area and will be greatly affected by solar radiation pressure. To counteract the effect of solar radiation pressure, the space sunshade needs to perform long-term orbit control, which will consume a large amount of propellant and further significantly increase the launch mass. Therefore, from the perspective of current aerospace technology capabilities, the space sunshade solution is not yet feasible.

From the perspective of material resource utilization, with the development of aerospace tech-

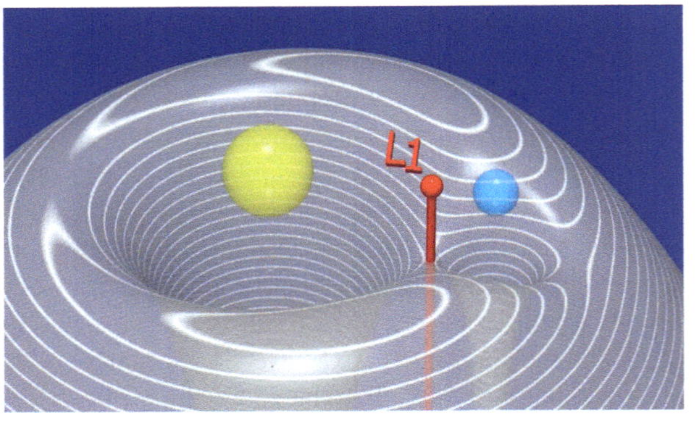

Fig. 1.16 Schematic diagram of Sun-Earth Lagrange point L1

Fig. 1.17 Large-scale thin-film spacecraft (Solar Sail). (Credit: NASA)

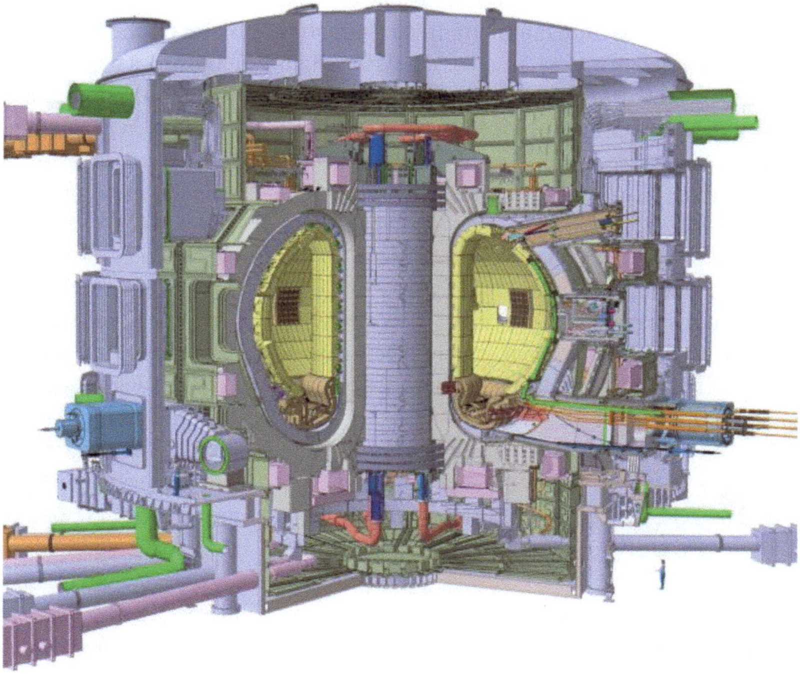

Fig. 1.18 Schematic diagram of International Thermonuclear Experimental Reactor (ITER)

nology, humans may be able to carry out large-scale resource development on extraterrestrial bodies. If we can extract and return important energy resources from outside the Earth for use, it will provide direct material support for solving global energy problems. At present, the most important extraterrestrial energy resource of concern is helium-3, which is abundant on the moon and is an important fuel for controlled nuclear fusion. The International Thermonuclear Experimental Reactor (ITER, Fig. 1.18) is a major scientific project jointly funded by the European Union, China, the United States, Russia, Japan, South Korea, and India. The goal is to build a superconducting Tokamak device that can produce large-scale nuclear fusion reactions, generate 500,000 kW of fusion power and sustain it for 500 s, with a total investment exceeding 10 billion US dollars. The current nuclear fusion mainly uses deuterium and tritium. Deuterium is extremely abundant in seawater and tritium can be regenerated in the reactor. But this reaction produces neutrons and has poor safety. The fusion of helium-3 and helium-3 does not produce neutrons and has excellent safety. The helium-3 near the Earth mainly comes from solar nuclear fusion and arrives in the earth's space through the solar wind. Due to the effects of the earth's magnetic field and atmosphere, the reserves of helium-3 on Earth are extremely small. The surface of the moon has been absorbing solar wind particles for 4 billion years, and the reserves of helium-3 are estimated to be as high as 1 million tons, so extracting helium-3 has become an important goal for future lunar resource utilization. At present, there is a great uncertainty in the commercial realization of controlled nuclear fusion, and the development and utilization of lunar helium-3 resources is also a very long-term plan, which cannot provide support for solving the global warming problem in a relatively long time.

From the perspective of energy resource utilization, the use of solar PV cells for power generation has been widely used on spacecraft. Because space solar energy resources are abundant and stable, if energy can be stably and efficiently transmitted to the ground, it will provide the Earth with an abundant, stable clean energy source. This will be the focus of this book, namely, space solar power.

1.3.2 The Concept of Space Solar Power

Space solar power (SSP), also known as space-based solar power (SBSP), refers to the system that convert solar energy into electric energy in space, and then transmit the energy wirelessly to the ground for terrestrial use. The concept of space solar power was first proposed by American Dr. Peter Glaser (Fig. 1.19) in 1968. The idea is to deploy solar power satellites in geostationary orbit (GEO), use a solar PV array with a diameter of about 6 km to receive sunlight and convert it into electricity, and then use a low-temperature superconducting power transmission system to transmit the electricity to a microwave transmitting antenna with a diameter of about 2 km. The antenna then continuously transmits energy to a ground receiving antenna with a diameter of about 3 km. Peter Glaser applied for a patent for this (Fig. 1.20) and published a paper in the Science journal.

An SSP system mainly consists of three parts: a solar power generation device, an energy conversion and transmission device, and a ground energy reception and conversion device (Fig. 1.21). The solar power generation device converts solar energy into electric energy; the energy conversion and transmission device converts electric energy into microwaves or lasers (lasers can also be directly converted from solar energy), and beam to the ground using a microwave antenna or optical system; the ground energy reception system uses a receiving antenna or laser PV array to receive the beams transmitted from space, and converts them into electric power through an energy conversion device on the ground. The entire process undergoes the energy conversion of solar energy-electricity-microwave (laser)-electricity, or solar energy-laser-electricity.

Fig. 1.19 Peter glaser, the proposer of solar power satellite

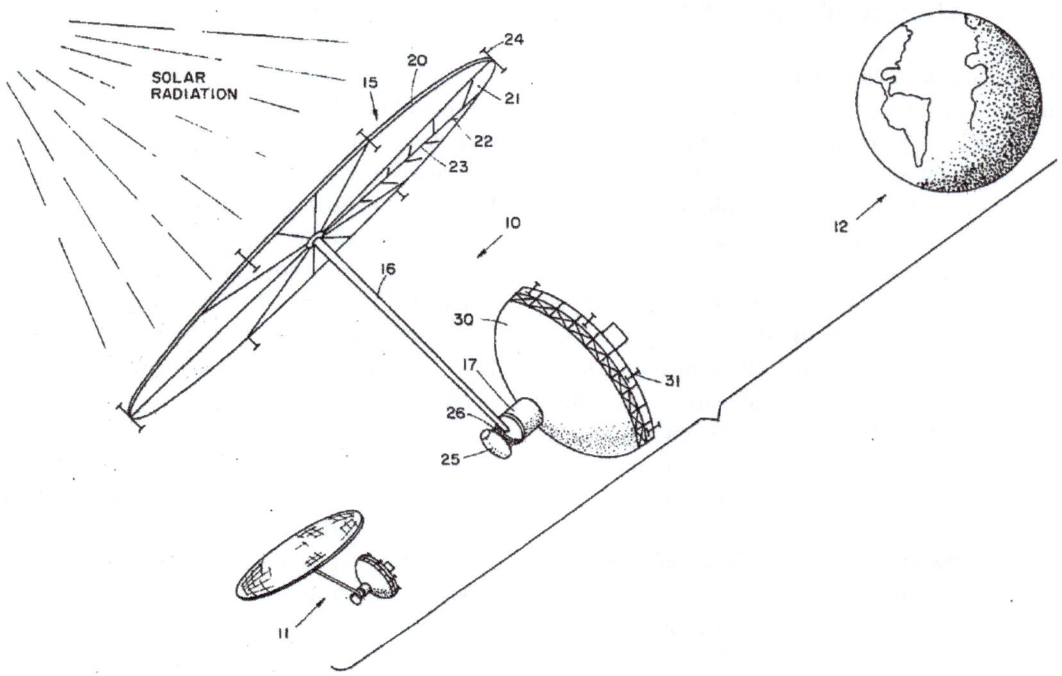

Fig. 1.20 Patent schematic diagram of solar power satellite (US Patent No. 5019768)

Fig. 1.21 Schematic diagram of an SSP system

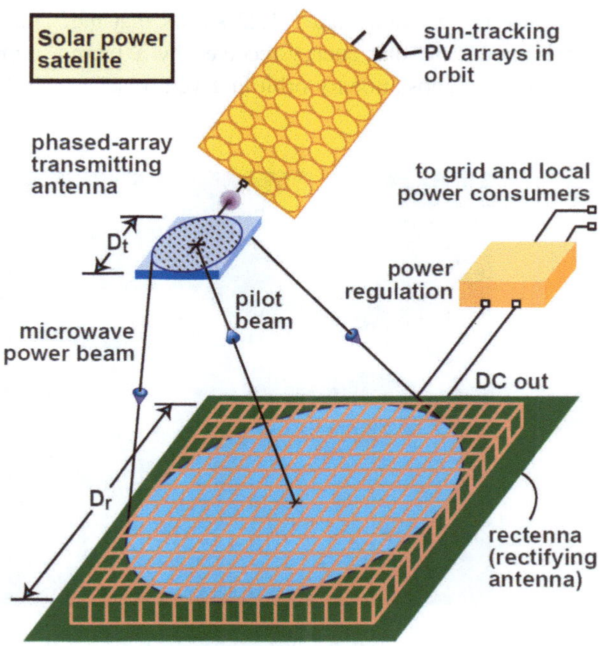

1.3.3 Characteristics of SSP

Compared to using solar energy on the ground, using solar energy in space has very prominent advantages. First, space solar energy resources are abundant and solar radiation energy is stable, so it is not affected by atmospheric attenuation, nor is it affected by seasonal and day-night changes. Second, for geostationary orbit, sunlight can be continuously received almost all year round and the utilization rate of sunlight exceeds 99%, so energy can be directly transmitted to the ground and provide stable large-scale clean energy without the need for huge energy storage facilities. SSP has become an important way to solve future energy and environment problems.

1.3.3.1 Abundant Space Solar Energy Resources

Because the solar radiation in space is higher than the solar radiation intensity on the ground, and the space available for solar energy utilization in space far exceeds that on the ground, the solar energy resources that can be utilized in space far exceed those on the ground. Figure 1.22 shows that the total amount of solar energy received by a 1 km wide ring in geostationary orbit in 1 year reaches 212 TW, which is almost equivalent to the total known oil reserves on Earth.

However, in practical applications, considering the orbital drift and control capabilities of an SPS, it is not possible to deploy lots of SPSs seamlessly in geostationary orbit. Assuming that the allowable drift range of an SPS is ±0.1°, i.e., one is arranged for every 0.2° (146.5 km), the total number of deployments is 1800. For an SSP system with a power of 1 GW, the total power generation can reach 1.8 billion kW. Based on this, the power generation capacity can be increased in the following ways:

1. Increase the power of an SSP system
2. Improve orbital control capabilities and reduce the drift range of an SPS
3. Use other orbits that can be used for relatively stable power transmission, such as inclined geosynchronous orbits

1.3.3.2 Stable Space Solar Radiation Intensity

The space is a vacuum environment, where the intensity of solar radiation remains constant, approximately 1360 W/m^2. However, the sunlight on the ground is affected by day and night,

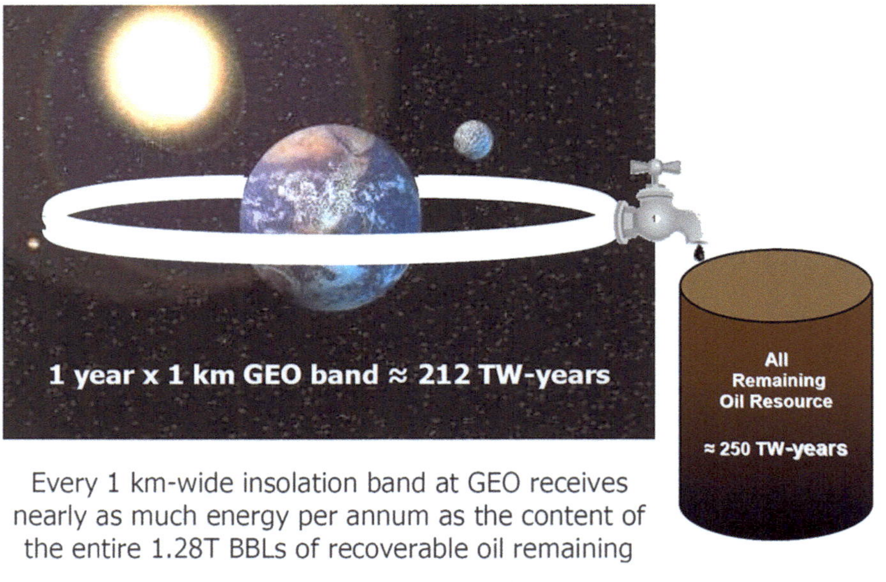

Fig. 1.22 Schematic diagram of solar energy resource on GEO. (Credit: DOD)

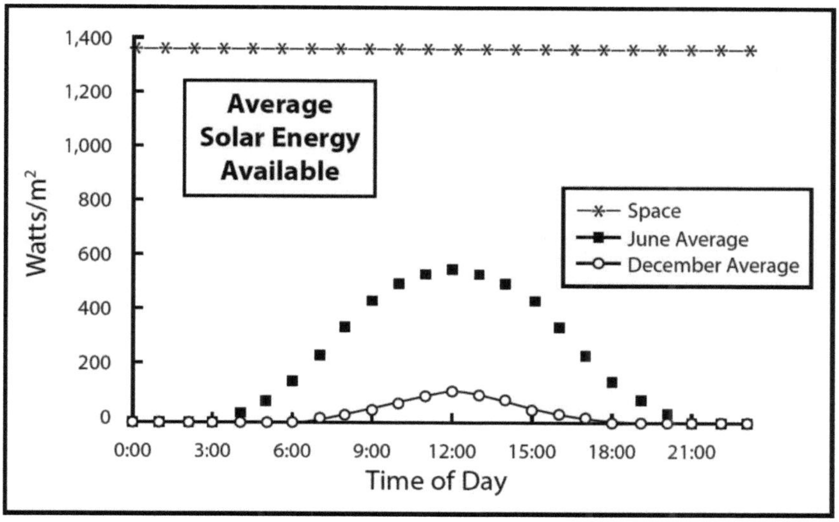

Fig. 1.23 Typical space and ground solar radiation intensities

Table 1.2 Classification of solar radiation intensity regions in China

Grade	Resource area lever	Total annual radiation/(MJ/m^2)	Total annual radiation/(kWh/m^2)	Average daily radiation/(kWh/m^2)	Ratio of space to ground
Richest area	I	≥6300	≥1750	≥4.8	≤6.83
Very rich area	II	5040–6300	1400–1750	3.8–4.8	6.83–8.63
Richer area	III	3780–5040	1050–1400	2.9–3.8	8.63–11.3
General	IV	<3780	<1050	<2.9	>11.3

and the intensity of solar radiation varies greatly, experiencing fluctuations from zero to maximum radiation intensity every day, as shown in Fig. 1.23 for typical space and ground solar radiation intensities. At the same time, the sunlight on the ground is affected by a combination of seasons, weather, latitude, etc., and the daily solar radiation varies greatly, with the overall average solar radiation intensity being relatively low. Taking the solar radiation intensity on the ground in China as an example (Table 1.2), for different light-exposed areas, the range of the total annual solar radiation intensity is 1050–1750 kWh/m^2, and the average radiation intensity is 120–200 W/m^2, which means that the average solar radiation intensity in space is 7–11 times of that on the ground.

1.3.3.3 No Need for Large-Scale Energy Storage Facilities

The difficulty of the ground solar power station as the main power supply system lies in the significant influence of day and night, weather, seasons, etc., with severe power fluctuations, especially unable to provide power during the night and rainy days. Without the support of other relatively stable large-scale power supply systems, it is necessary to equip with massive energy storage facilities to provide stable power supply. The configuration should be based on relatively harsh conditions, and the development of energy storage technology has become the core issue restricting the large-scale application of ground solar power stations as the main power supply system. Figure 1.24 shows that after considering

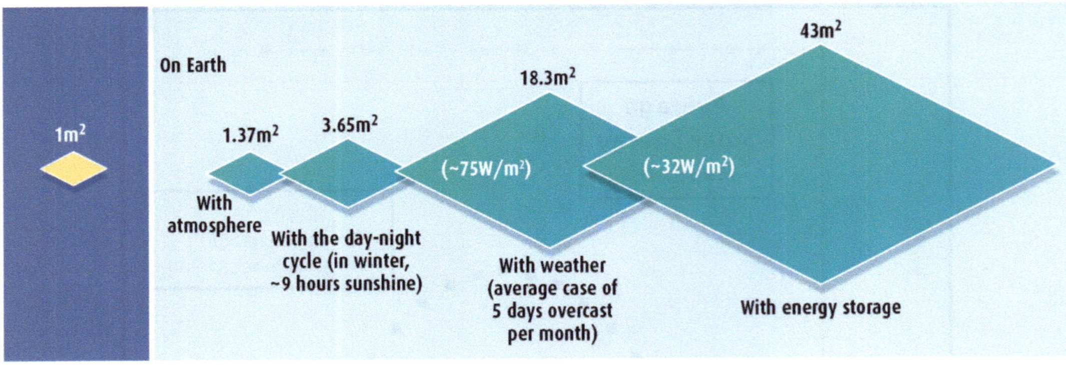

Fig. 1.24 PV area requirement considering various factors for solar power station on the ground. (Credit: NewScientist)

various factors such as day and night, seasons, continuous 5 days of rainy weather, and energy storage efficiency, the area of the solar PV required for the ground solar power station to achieve stable power supply with energy storage facilities will reach 43 times the space (not considering power transmission efficiency). However, an SPS operates in geostationary orbit, only undergoing up to 72 min of Earth shadow period each day for 6 weeks around the spring and autumn equinoxes each year, and can achieve continuous stable illumination at other times. If the SPS can achieve solar PV array Sun-pointing and power transmission device Earth-pointing, it can almost achieve continuous power generation 365 days a year, 24 h a day, without the need to equip massive energy storage facilities in space or on the ground, making it suitable as the main power supply system.

However, the scale of SPS is enormous, and the construction is extremely difficulty. It is one of the most magnificent space super-projects conceived by humans by now. The current ability of space technology cannot support the construction of SPS, and the technical feasibility is a key factor restricting the development of SSP. In addition, if we analyze according to the current cost of spacecraft development, launch, and operation, the power generation cost of SSP will be extremely high. Economic feasibility is another key factor restricting the commercial development of SSP. The adoption of new technologies, new schemes, and new aerospace development models to reduce the overall cost of SSP is crucial for future large-scale commercial applications. At the same time, the safety of the actual development, construction, and operation of SSP is a matter of great concern to the public. The potential impact of long-term high-power energy transmission on the environment and the safety impact on other existing facilities need to be fully studied.

1.4 International Development of SSP

1.4.1 United States

The United States is the earliest and largest investor in the field of SSP. In recent years, departments mainly composed of the California Institute of Technology (Caltech), the US Naval Research Laboratory (NRL), and the US Air Force Research Laboratory (AFRL) have gradually increased their R&D efforts in this field and initiated a new round of large-scale research.

In 2013, Caltech received a hundred of millions of dollars in private capital funding to conduct research on new structures, efficient power generation, and wireless power transmission of SSP. In 2015, it also got a $17.5 million R&D project from Northrop Grumman Corporation, proposed the concept of "Microwave Swarm" SSP, and conducted key technology space verification in early 2023. NRL has carried out continuous technical research and verification work in the field of microwave and laser wireless

Fig. 1.25 PRAM test apparatus on X-37B. (Credit: NRL/NASA)

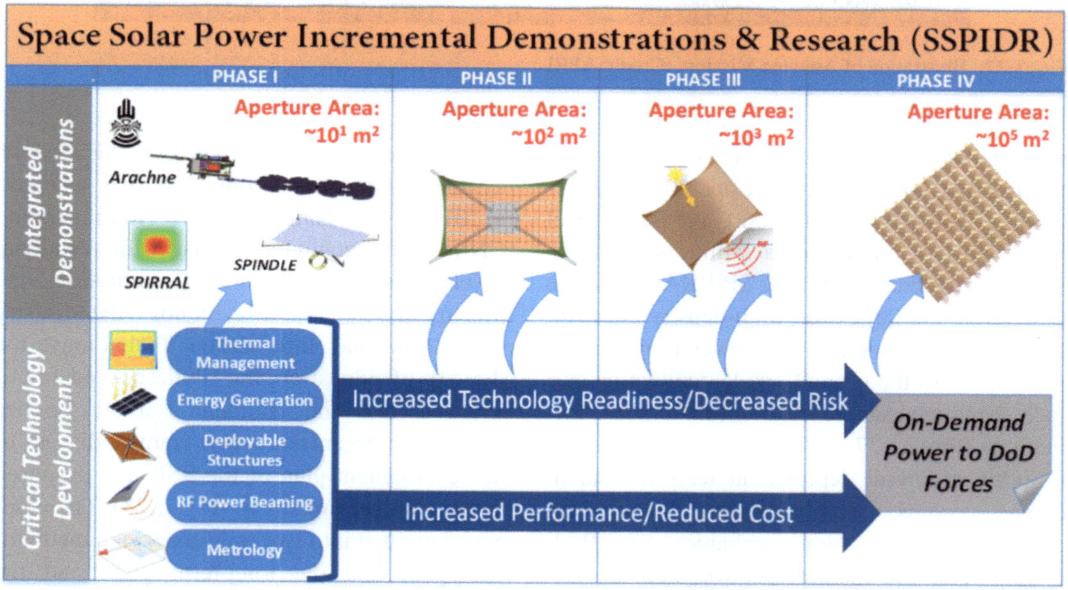

Fig. 1.26 Space Solar Power Incremental Demonstrations and Research (SSPIDR) program. (Credit: AFRL)

power transmission technologies. Its microwave power transmission module (PRAM) was tested on the X-37B in May 2020 (Fig. 1.25). PRAM is a sandwich-structure module of 30 cm × 30 cm, including photovoltaic cells, DC-RF converters, and transmitting antennas. Its main function is to collect sunlight through solar PVs and convert it into electric power, and then convert the electric power into microwaves for transmission. In 2021, NRL conducted a ground test verification of kilometer-level transmission of kilowatt-level microwave energy.

In November 2019, AFRL obtained a $100 million R&D contract with Northrop Grumman Corporation to launch the Space Solar Power Incremental Demonstrations and Research (SSPIDR) program (Fig. 1.26). The main purpose is to develop extremely lightweight, ultra-thin, flexible, highly efficient packaging, on-orbit deployment, ultra-large-scale solar-to-microwave conversion units, and verify the technical feasibility of rapid low-cost power supply to the ground. The SSPIDR program includes four verification stages: the first stage, principle component prototype verification, is at the 10 m^2 size level; the second stage, scaled integrated prototype verification, is at the 100 m^2 size level; the third stage, full-size SPS spacecraft prototype

Fig. 1.27 Illustration of Arachne Mission. (Credit: AFRL)

verification, is at the 1000 m^2 size level; the fourth stage, completion of the operating constellation, achieves the 100,000 m^2 size level SPS. The first phase includes three verification projects: Arachne, SPINDLE, SPIRRAL. Arachne will verify the large-size, high-density encapsulated low-cost modules for power transmission from space to the ground, while conducting surface shape measurements to optimize beam formation (Fig. 1.27). SPINDLE will conduct on-orbit structural deployment tests of scaled-down systems, mainly testing deployment dynamics and structural mechanics. SPIRRAL mainly tests the thermal control methods that ensure the system's high-performance operation. At the same time, NRL is responsible for developing extremely low power intensity (1–1000 mW/m^2) receiving and rectifying technologies.

1.4.2 China

Since the "Eleventh Five-Year Plan," China has continuously carried out research in the field of SSP, focusing on the system design and key technologies such as large-scale space solar energy collection and conversion, microwave wireless power transmission, and laser wireless power transmission. In 2014, the China Academy of Space Technology proposed a new type of MR-SPS, and Xidian University proposed a new type of concentrated SSPS-OMEGA SPS, which became internationally representative schemes. In 2014, China National Space Administration (CNSA), in conjunction with the National Development and Reform Commission, the Ministry of Science and Technology, the Ministry of Industry and Information Technology, the Ministry of Education, the Chinese Academy of Sciences, and the National Natural Science Foundation, organized expert team to discuss of the development plan of the SSP, and proposed the development roadmap (Fig. 1.28). The roadmap is divided into two milestones, the medium-term goal and the long-term goal, and the medium-term goal is further divided into three stages every 5 years.

1. Medium-term goal
   By 2030, start the construction of an MW-level demonstration SSP, achieve emergency power supply verification and scientific research.
2. Long-term goal
   By 2050, has the ability to build a GW-level commercial SSP, which can provide abundant renewable clean energy, and meet the strategic needs of sustainable development.

On December 6, 2018, the construction of the SSP Experiment Base in Bishan District, Chongqing was launched. The project covers an area of about 200 acres (about 13.33 ha), with an investment of 100 million yuan, mainly for long-

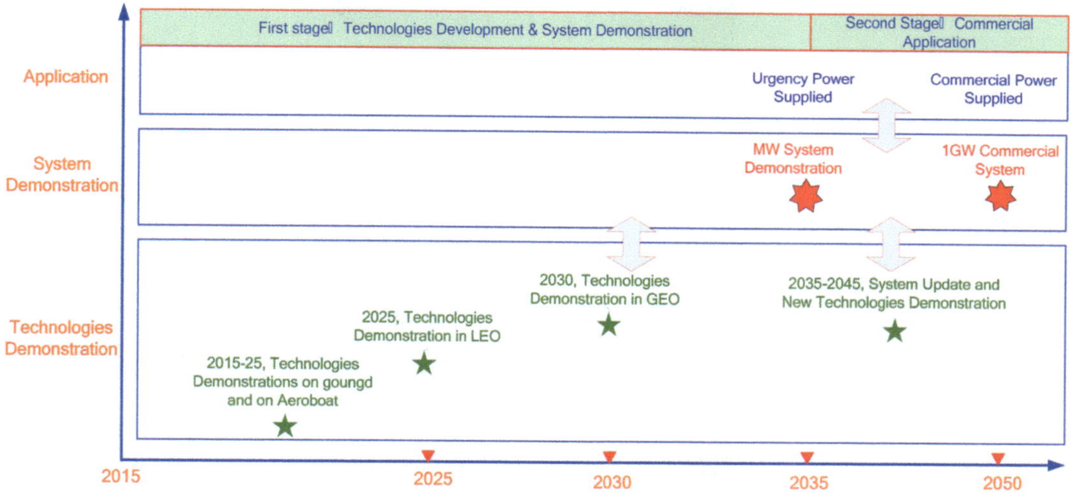

Fig. 1.28 Proposed roadmap of SSP in China

Fig. 1.29 Bishan SSP experiment base. (Credit: Chongqing University)

term experimental research in wireless power transmission, high-voltage high-power electrical systems, power transmission and environmental impact, and supports large-scale SSP system research (Fig. 1.29). On December 23, 2018, the "Chasing the Sun Project" was launched at Xidian University. The project is based on the SSPS-OMEGA concept and builds a full-system, full-link SSP ground verification center to carry out technical verification of the entire SSP system on the ground (Fig. 1.30).

1.4.3 Japan

Japan started research in the field of SSP in the 1980s, and officially included the development of SSP in the Basic Space Plan in 2009. With the

Fig. 1.30 SSP ground verification center. (Credit: Xidian University)

support of the Ministry of Economy, Trade and Industry (METI) and the Ministry of Education, Culture, Sports, Science and Technology (MEXT), the Japan Aerospace Exploration Agency (JAXA), and Japan Space Systems (JSS) have jointly carried out long-term research on wireless power transmission technology and experiments with many companies and universities. In 2015, Japan successfully carried out a landmark kilowatt-level high-precision microwave wireless power transmission experiment on the ground, and in 2019 successfully verified the microwave wireless energy supply technology for an unmanned aerial vehicle. According to the updated development roadmap in 2016 (Fig. 1.31), in 2023–2025, a kilometer-level vertical wireless power transmission (WPT) technology verification will be carried out; by 2030, a WPT technology verification from low orbit to the ground will be carried out; a WPT technology verification from GEO to the ground by 2035 will be carried out; by 2040, a technical verification of the basic modules of the GW SPS will be carried out. In 2021, Japan proposed in its revised Basic Space Plan to carry out a low-orbit verification mission called DELIGHT before 2025 (Fig. 1.32), mainly to verify the deployment characteristics of flat antennas and to carry out the microwave power transmission test between space and ground.

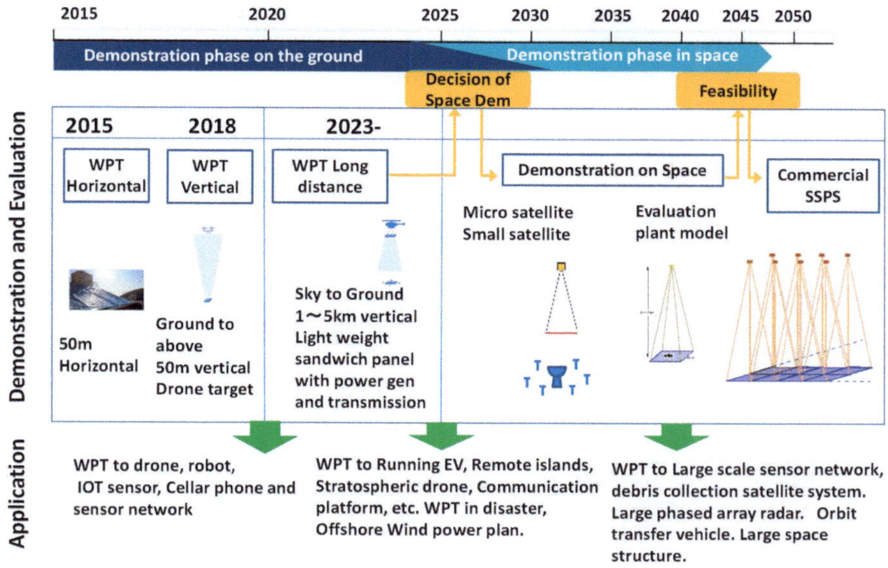

Fig. 1.31 Japan's roadmap for SSP in 2016. (Credit: Japan Space Systems)

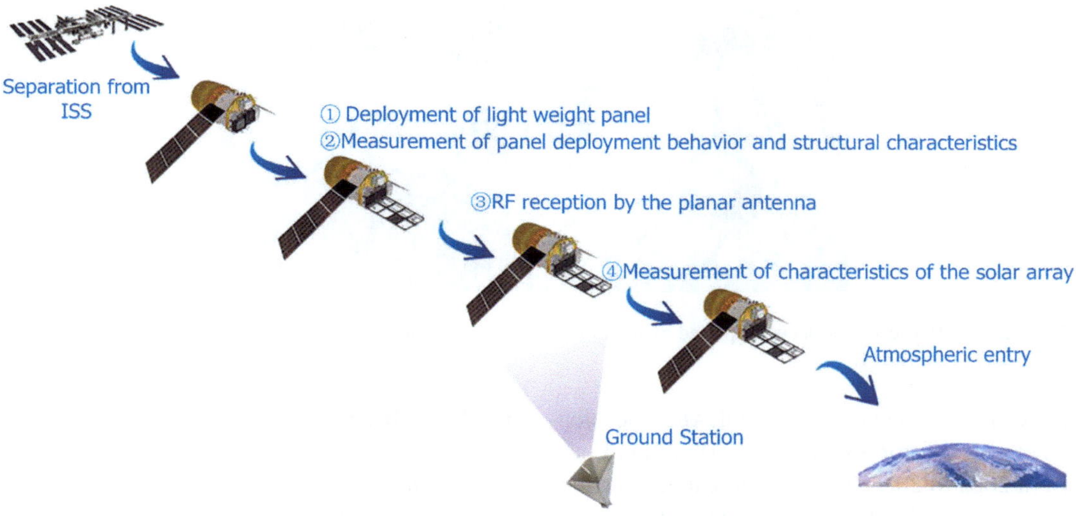

Fig. 1.32 Scheme of low-orbit verification mission DELIGHT. (Credit: JAXA)

1.4.4 European Space Agency (ESA)

In March 2021, ESA officially released the "ESA AGENDA 2025," which identified the top five priority development directions for ESA before 2025. The second priority development direction is to boosting commercialization for a green Europe, clearly stating that space technology will support a circular economy, climate neutrality and energy transition in Europe, and potential space-based solar power generation deserve to be further investigated. In 2021, ESA restarted the research on SSP, supporting 13 project studies in 8 directions (Fig. 1.33). At the beginning of 2022, ESA commissioned the Frazer-Nash and the Roland Berger to conduct the independent economic assessment, and officially released the assessment report in August. In January 2023,

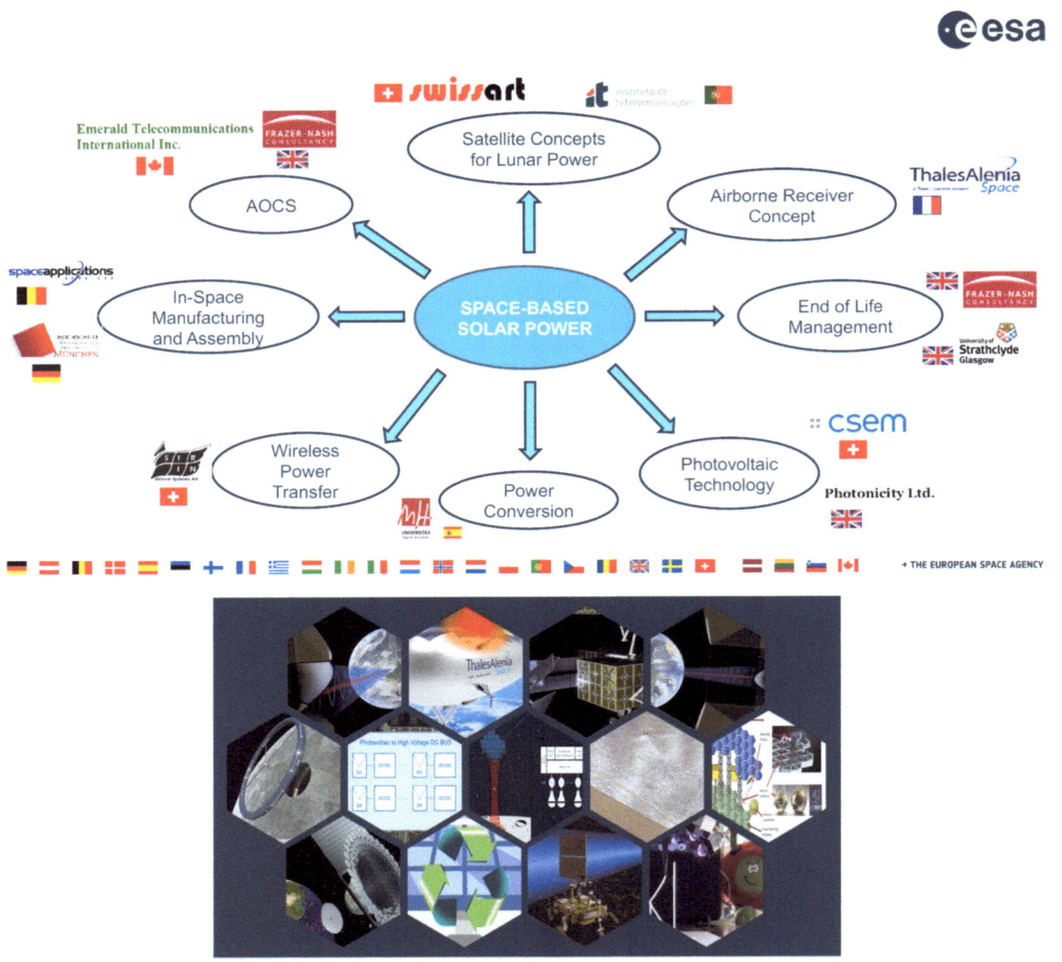

Fig. 1.33 SSP research projects supported by the European Space Agency. (Credit: ESA)

ESA officially launched the SOLARIS development program for SSP, funding 60 million Euros from 2023 to 2025. The main research directions are shown in Fig. 1.34, and the possible development roadmap is shown in Fig. 1.35. It is planned to complete the construction of a 1–100 MW demonstration SPS before 2035, to achieve the operation of commercial SPS before 2040, and to support Europe's goal of net zero carbon emissions by 2050.

1.4.5 Other Countries

South Korea established the space solar power research association in 2016, aiming to serve future net zero emission targets and promote the development of SSP. In 2018, the research project on SSP was officially launched, which is led by the Korean Aerospace Research Institute (KARI) and the Korean Electric Power Research Institute (KERI), and proposed South Korea's

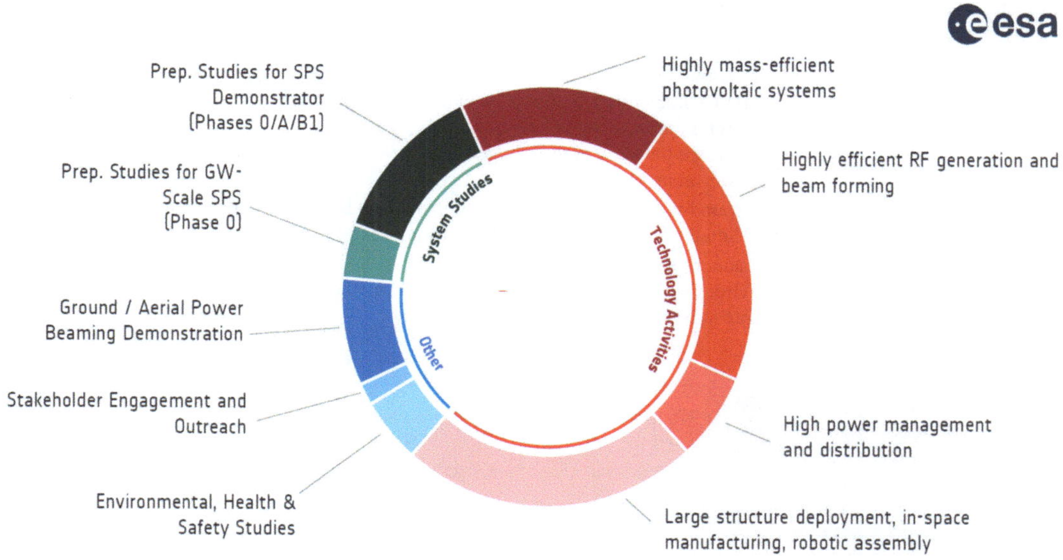

Fig. 1.34 Research directions of SOLARIS program. (Credit: ESA)

Fig. 1.35 Proposed roadmap of SSP by ESA. (Credit: ESA)

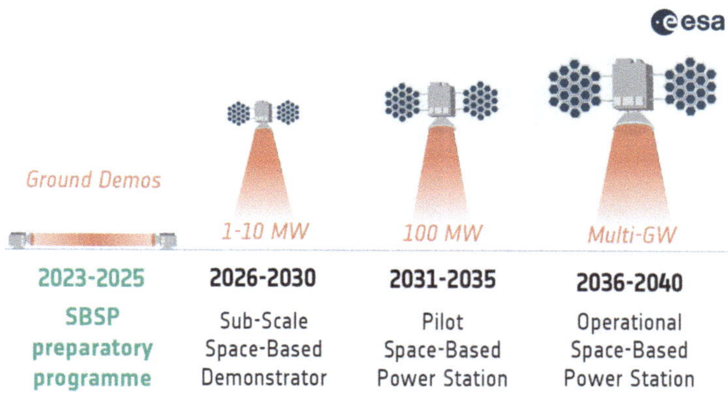

plan to develop SSP, focusing on system, structure, high-voltage power, and microwave wireless power transmission technology research. It plans to carry out ground tests of inter-satellite wireless power transmission technology from 2023 to 2026, and launch two small satellites for wireless power transmission technology demonstration from 2027 to 2029. The South Korean development plan of SSP from 2020 to 2029 is shown in Table 1.3.

In 2020, the UK Department for Business, Energy and Industrial Strategy (BEIS) commissioned Frazer-Nash Consultancy to assess the prospects for SSP. In September 2021, they officially released the report "Space Based Solar Power: De-risking the pathway to Net Zero," proposing SSP as a new option for the UK to achieve net zero, and suggesting a development roadmap (Fig. 1.36). It includes ground verification and balloon tests from 2022 to 2026, 40 MW SPS demonstrator in low Earth orbit from 2027 to 2031, 500 MW SPS demonstrator in operational orbit from 2032 to 2035, construction of a geostationary orbit 2 GW SPS prototype system from 2036 to 2039, and multiple SSP systems providing 15% of the UK's energy from 2040 to 2042. In September 2021, the UK Space Agency officially released the "National Space Strategy,"

Table 1.3 Development plan of SSP from 2020 to 2029 in South Korean

Schedule	Development contents
1st phase (2020—2022)	• Roll-out solar array and folded antenna • High-voltage/high-power management system • Wireless power transmission system • Interface between high-voltage/high-power management system and wireless power transmission
2nd phase (2023—2026)	• Ground test for wireless power transmission between small or cube satellites • 2 Small or cube satellites and environment test for SPS
3rd phase (2027—2029)	• Launch of 2 small or cube satellites for SPS Demo. in space • Wireless power transmission between 2 small or cube satellites in space and transmission efficiency measurement

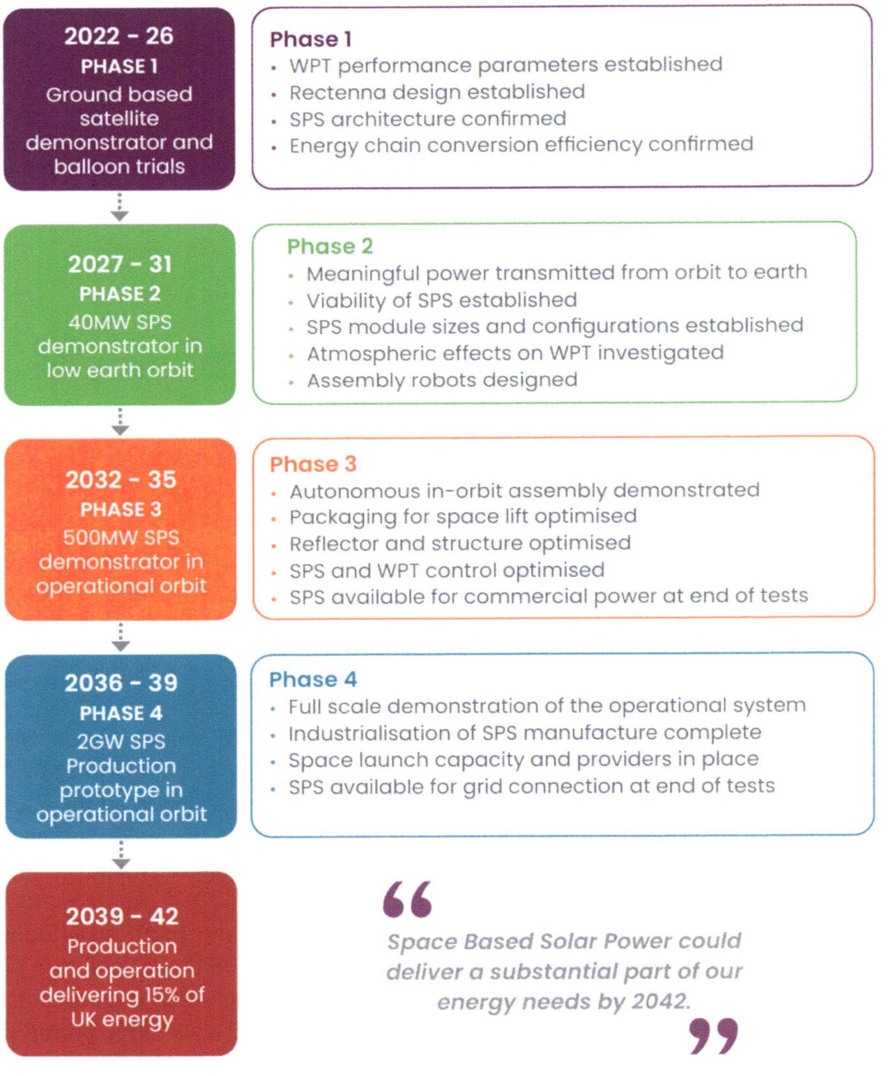

Fig. 1.36 Recommended SSP development roadmap in UK. (Credit: UK SEI)

Fig. 1.37 UK space energy initiative and participating organizations. (Credit: UK SEI)

proposing the development of SSP as a potential net zero energy source for the UK. At the same time, BEIS included SSP in the Net Zero Innovation Portfolio, with an initial investment of three million pounds. In March 2022, the UK officially launched the Space Energy Initiative (SEI), which includes dozens of related aerospace, energy companies, research institutes, and universities (Fig. 1.37), jointly promoting the development of the UK's SSP.

Bibliography

1. Glaser P E. Power from the Sun: Its Future[J]. Science, 1968, 162: 867-886.
2. ESA Agenda 2025[R]. ESA, 2021.
3. National Space Strategy[R]. UK Space Agency, 2021.
4. Net Zero by 2050: A Roadmap for the Global Energy Sector[R]. IEA, 2021.
5. The Space Energy Initiative[R]. The Space Energy Initiative, UK, 2021.
6. Space Based Solar Power: De-risking the pathway to Net Zero[R]. Frazer-Nash Consultancy, 2021.
7. Space-Based Solar Power: A Future Source of Energy for Europe[R]. Frazer-Nash Consultancy, 2022.
8. Space-Based Solar Power: Can It Help to Decarbonize Europe and Make It More Energy Resilient[R]. Roland Berger, 2022.
9. Catching the Sun: A National Strategy for Space solar power[R]. The Beyond Earth Institute, 2021.
10. Outline of the Basic Plan on Space Policy[R]. National Space Policy Secretariat, Cabinet Office, Japan, 2020.
11. Jaffe P, Borders K, Browne C, et al. Opportunities and Challenges for Space Solar for Remote Installations[R]. NRL/MR/8243--19-9813. Naval Research Laboratory, Washington, D. C., USA, 2019.
12. Mankins J C. SPS-ALPHA: The First Practical Solar Power Satellite via Arbitrarily Large Phased Array Final Report[R]. NASA NIAC, 2012.
13. Vedda J A, Jones K L. Space-Based Solar Power: A Near-Term Investment Decision[R]. Center for Space Policy and Strategy. Space Agenda 2021, 2020.
14. SPACE: The Dawn of a New Age[R]. Citi GPS: Global Perspectives & Solutions, 2022.
15. National Security Space Office. Space-Based Solar Power as An Opportunity for Strategic Security[R]. DOD, 2007.
16. Mankins J C. Fifty Years of Space solar power[C]. 69th International Astronautical Congress, Bremen, Germany, 2018.
17. Mihara S, Machida H, Sasaki K. Current Status of the SSPS Development and the Result of Ground to Air Microwave Power Transmission Experiment[C]. 70th International Astronautical Congress, Washington D. C., USA, 2019.
18. Mankins J C. IAA Decadal Assessment of Space solar power: A Progress Report[C]. 71th International Astronautical Congress, Dubai, UAE, 2021.
19. Summerer L, Vijendran S, Makaya A, et al. Space-Based Solar Power Plants - Outcome of A Thorough Cost Benefit Analysis in the Light of Achieving the Net-Zero CO2 Target by 2050[C]. 72th International Astronautical Congress, Paris, France, 2022.
20. Soltau M, Homfray D, Cash I, et al. The UK Space Energy Initiative – Towards a Practical Space Based Power System for the Net Zero Era[C]. 72th International Astronautical Congress, Paris, France, 2022.

21. Sasaki K, Machida H, Ijichi K, et al. The Outline and the Current Status of the Power Transmission System Development Project for the Realization of the SSPS[C]. 72th International Astronautical Congress, Paris, France, 2022.
22. Choi J M, Yi S H. Proposal of the First Korean Pilot System for Space Based Solar Power (SBSP) [C]. 72th International Astronautical Congress, Paris, France, 2022.
23. Mihara S, Maekawa K, Nakamura S, et al. The Current Status of Microwave Power Transmission for SSPS and Industry Application[C]. 68th International Astronautical Congress, Adelaide, Australia, 2017.
24. Mihara S, Maekawa K, Nakamura S, et al. The Road Map toward the SSPS Realization and Application of Its Technology[C]. 69th International Astronautical Congress, Bremen, Germany, 2018.
25. Hoffert M I, Caldeira K, Benford G, et al. Advanced Technology Paths to Global Climate Stability: Energy for A Greenhouse Planet[J]. Science, 2002, 298: 981-987.
26. Retherford C. The Promise of Space-Based Solar Power[J]. Space Policy Review, 2022(1): 1-24.
27. Mankins J C. The Case for Space solar power[M]. Houston, TX: Virginia Edition Publishing, 2014.
28. Zhuang F G, Li M, Wang L, Hou X B, et al. Solar Power Satellite: Strategic Combination of Future Space Technology and New Energy Sources[J](in Chinese). Aerospace China, 2008(7):36-39.
29. Wang X J, Min G R. New Technology Industrial Revolution Triggered by the Development of Space Solar Power Station[N] (in Chinese). Science Times, 2011-12-7(A1).
30. Summary for Policymakers of the 2014 Climate Change Synthesis Report[R]. IPCC, 2014.
31. Summary for Policymakers of the Global Warming of 1.5°C[R] (in Chinese). IPCC, 2019.
32. Jaffe P. A Feasibility Assessment for Providing Energy to Remote Installations via Space Solar[C]. 70th International Astronautical Congress, Washington D. C., USA, 2019.
33. https://www.esa.int/Enabling_Support/Space_Engineering_Technology/SOLARIS/
34. https://physicsworld.com/a/space-based-solar-power-could-beaming-sunlight-back-to-earth-meet-our-energy-needs/
35. https://www.spacedevelopmentfoundation.org/wordpress/space-solar-power-symposium-2022-agenda/

Design Basics of Space Solar Power

2

2.1 Operational Orbit Characteristics

The operational orbit greatly affects the operation and the scale of SPS and the reception of energy on the ground. Factors that need to be considered include the illumination conditions of the solar array, the altitude of the operational orbit, and the relative position between the SPS and the ground receiving station. The illumination conditions of the solar array determine whether the SPS can generate electricity continuously; the altitude of the orbit determines the distance of wireless power transmission, directly affecting the size of the transmitting and receiving antennas, as well as the cost of transportation; the relative position between the SPS and the ground receiving station determines the continuity of wireless power transmission and the range of beam steering angles.

Currently, the main operational orbits considered for SPSs include low Earth equatorial orbit, low Earth inclined orbit, Sun-synchronous orbit (Fig. 2.1), geostationary orbit, geosynchronous orbit (Fig. 2.2), Sun-Earth L1 point orbit, lunar orbit, etc.

1. Low Earth equatorial orbit (LEEO)
   LEEO refers to a circular orbit with an inclination of 0° located above the equator with an altitude of 200–1000 km. The main advantages of the LEEO include: lower orbit altitude, lower transportation cost, convenient for on-orbit assembly and maintenance, shorter wireless power transmission distance, smaller transmitting and receiving antenna area, only setting up ground receiving stations near the equator. The disadvantages of LEEO include: a longer Earth shadow period, during which the solar energy cannot be achieved; the relative position between the SPS and the receiving antenna is constantly changing, the time of WPT to a receiving station is short; if continuous power supply is to be ensured, multiple SPSs need to be launched on the orbit to form a constellation and multiple receiving stations need to be deployed; due to the limitation of the beam steering angle, the receiving station can only be located near the equator. This orbit is suitable for power supply in areas near the equator and also suitable for verification of key technologies and small system-level demonstration.
2. Low Earth inclined orbit (LEIO)
   LEIO refers to the circular orbit with an orbital inclination of not 0° and an orbit altitude of 200–1000 km. The main advantages of LEIO include: lower orbit altitude, lower transportation cost, convenient for on-orbit assembly and maintenance; shorter wireless power transmission distance, smaller transmitting and receiving antenna area; ground receiving stations can be arranged over a larger range. The disadvantages of LEIO include: a longer shadow period, during which power

X. Hou, *Space Solar Power*, Sustainable Development Goals Series,
https://doi.org/10.1007/978-981-97-3580-8_2

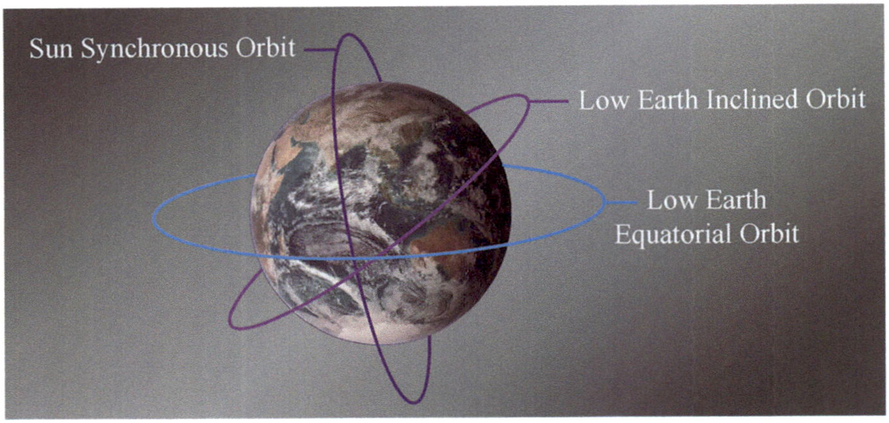

Fig. 2.1 Schematic diagram of low Earth equatorial orbit, low Earth inclined orbit and Sun-synchronous orbit

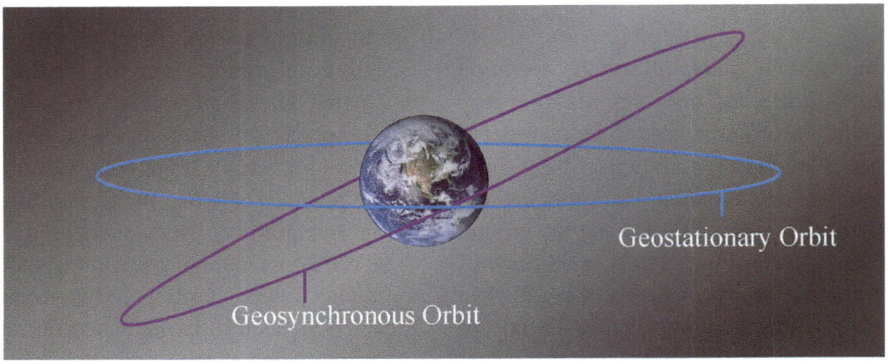

Fig. 2.2 Schematic diagram of geosynchronous orbit and geostationary orbit

transmission cannot be achieved; the relative position between the SPS and the receiving antenna is constantly changing, the time passing through the receiving station is short; if continuous power supply is to be ensured, multiple SPSs need to operate on multiple orbits to form a constellation and multiple receiving stations need to be deployed on the ground, and a larger beam scanning angle is required; the solar array needs two-dimensional rotation to orient the sun and the transmitting antenna pointing at the earth. This orbit is suitable for establishing a global SPS constellation for global power supply, but the corresponding number of SPSs and ground receiving stations is large.

3. Sun-synchronous orbit (SSO)

   SSO refers to the orbit plane's rotation direction and angular velocity around the earth being consistent with the earth's rotation direction and angular velocity around the sun that means the orbit plane always maintains a relatively fixed direction with the sun. The corresponding orbit inclination is close to 90° and the orbit altitude is from several hundred kilometers to several thousand kilometers. SSO often used for remote sensing satellites. The main advantages of SSO include: lower orbit altitude, lower transportation cost, convenient for on-orbit assembly and maintenance; shorter wireless power transmission distance; therefore, the transmitting and receiving antenna area is smaller; for a specific morning and evening SSO, the sun orientation of the solar array and the earth-pointing of the transmitting antenna can be well maintained, there is no need for relative rotation, and continuous power generation can be achieved for most of the year. The disadvantages of SSO include: the relative change in position between the SPS and the receiving antenna, the short time passing through the receiving station, if continuous power transmission is to be ensured, a large number of SPSs need to be arranged on multiple SSOs to form a complex constellation, and a larger beam scanning angle is required; to better receive power, a large number of receiving stations need to be arranged globally. This orbit is more suitable for key technology verification and small system-level demonstration and can also be used for discontinuous power supply in the South Pole and North Pole.

4. Geostationary orbit (GEO)

GEO refers to a circular orbit located above the equator, where the orbital period is exactly the same as the earth's rotation period (23 h, 56 min, and 4 s) and direction, meaning that the orbit remains relatively stationary with the earth, i.e., the sub-satellite point is unchanged. The corresponding orbital inclination is 0°, the orbit altitude is 35,786 km, and it is commonly used for communication satellites. GEO is the best operating orbit for SPSs, as it can ensure continuous sunlight for the solar array for most of the year, with only up to 72 min of shadow period per day for about 6 weeks around the spring and autumn equinoxes (Fig. 2.3). It is easy to achieve fixed-point continuous power transmission between the transmitting antenna and the ground receiving antenna without the need for large-scale beam steering. The disadvantages of GEO are: high orbit altitude, high transportation cost, difficulty in on-orbit assembly and maintenance; far from the earth, long power transmission distance, therefore large transmitting and receiving antenna area; the solar array and the transmitting antenna maintain relative rotation to ensure the solar array pointing towards the sun and the transmitting antenna pointing at the earth.

5. Geosynchronous orbit (GSO)

GSO refers to a circular orbit where the orbital period is the same as the earth's rotation period, and GEO is a special type of GSO. For GSOs with an inclination not equal to 0°, the sub-satellite point position will change slightly, presenting an "8" shape. GSO is the second-best operating orbit for SPSs after GEO, with the main features similar to GEO, but requiring a larger beam steering range (related to the orbital inclination). The application of GSO can increase the number of deployable SPSs, which is of great value for expanding the overall power supply capacity of SSP.

6. Sun-Earth L1 orbit

The libration point, also known as the Lagrange point, is one of the five special solutions in the restricted three-body problem of celestial mechanics, corresponding to the five spatial positions where a small object remains stationary relative to two large objects (Fig. 2.4). For the Sun-Earth Lagrange points, L1 and L2 are 1.5 million km from the Earth, and L3 and L4 are 150 million km from the Earth. The Sun-Earth L1 orbit, due to its special position, is suitable for deploying SPSs. The advantages of the Sun-Earth L1 orbit include: easy to achieve sun orientation of the solar array and earth orientation of the transmitting antenna, no need for mutual rotation between the solar array and the transmitting antenna; continuous power transmission can be ensured simply by steering the beam direction; the operating orbit has no shadow period and can generate electricity continuously; the orbit space is large, allowing for the arrangement of a large number of SPSs. The disadvantages of the Sun-Earth L1 orbit include: far from the earth, high transportation cost, difficult assembly and maintenance; long power transmission distance, large transmitting and receiving antenna area; due to the earth's rotation, multiple receiving stations need to be set up on the ground to receive energy continuously.

Fig. 2.3 Shadow period of GEO

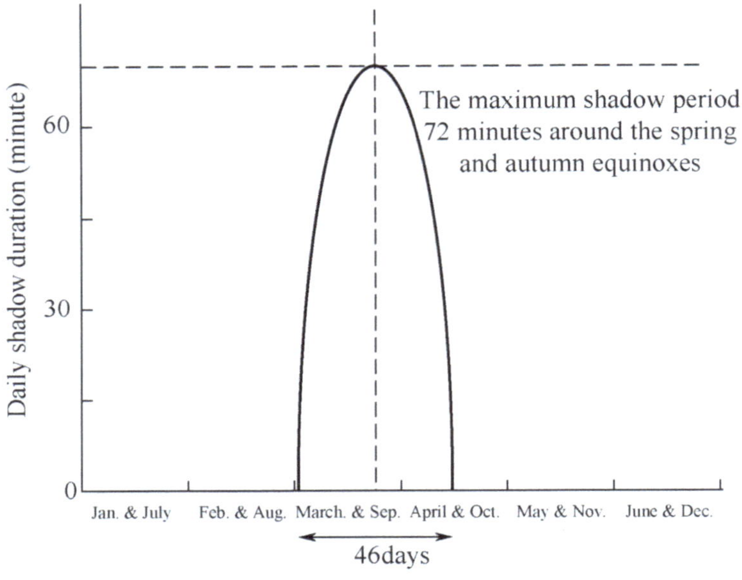

Fig. 2.4 Earth-Sun lagrange libration points

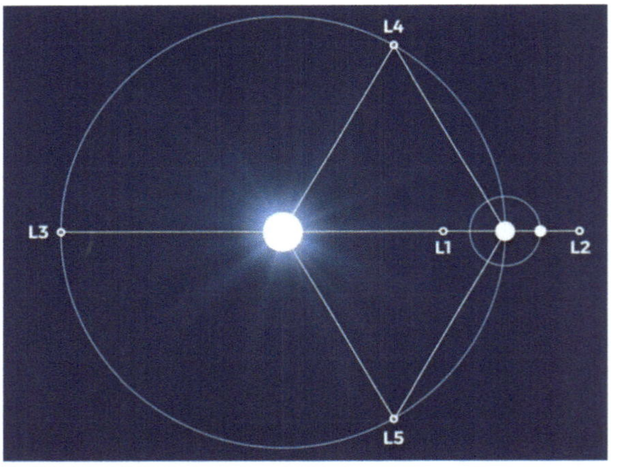

7. Lunar orbits

   The lunar orbits mainly refer to the circumlunar orbit, Earth-Moon L1 orbit, and Earth-Moon L2 orbit (about 60,000 km from the moon), which can be used to power the facilities on the lunar surface. Due to the moon's special long eclipse period of 14 days and the existence of some permanent shadow areas on the lunar surface, the energy supply for lunar exploration and resource utilization is extremely challenging. If a laser power transmission SPS is deployed in the circumlunar orbit, Earth-Moon L1 point orbit, or Earth-Moon L2 point orbit, it can provide continuous or non-continuous power supply for lunar rovers or lunar bases and may become an important power supply method for future lunar exploration.

2.2 Characteristics of Space Environment

2.2.1 Space Environment

The design life of an SPS is expected to exceed 30 years. The main factors of the space environment, including vacuum and cold environment, microgravity environment, solar electromagnetic radiation environment, atomic oxygen environment, space plasma environment, space charged particle radiation environment, space debris and micrometeoroid environment, etc. (Fig. 2.5), have a significant impact on the life of an SPS. The SPS will experience orbital changes from LEO to GEO, and the space environment on different orbits has its own characteristics, and its impact is also different.

1. Vacuum environment
   The outer space is a high vacuum environment, the higher the orbit, the higher the vacuum. At 600 km, the atmospheric pressure drops to below 10^{-7} Pa, and at 10,000 km, the atmospheric pressure is about 10^{-10} Pa.
2. Cold environment
   Without considering the heat radiation of nearby celestial bodies, the entire space can be regarded as an ideal black body with extremely low temperature. The cold environment mainly refers to the 3K background radiation environment in space, which is equivalent to a heat sink environment close to absolute zero.
3. Microgravity environment
   The spacecraft runs on a space orbit at a certain speed, and the gravity and centrifugal force it receives are balanced. However, due to the existence of a certain non-circular orbit and space disturbance forces, the spacecraft is actually in a microgravity state, and the corresponding gravity acceleration is (10^{-6}–10^{-3}) g.
4. Solar electromagnetic radiation environment
   The sun is a huge radiation source, transmitting wavelengths from 10^{-14} m γ rays to 10^2 m radio waves, with the main energy concentrated in the far ultraviolet, near ultraviolet, visible light, and infrared bands. According to the American ASTM 490 standard, the solar electromagnetic radiation power intensity at a distance of one astronomical unit from the sun is about 1353 W/m^2, it is also known as the solar constant.
5. Atomic oxygen environment
   Atomic oxygen is the residual gas existing in atomic oxygen state on low Earth orbits (200–700 km), mainly caused by the photolysis of oxygen molecules in the atmosphere by ultraviolet rays with wavelengths less than

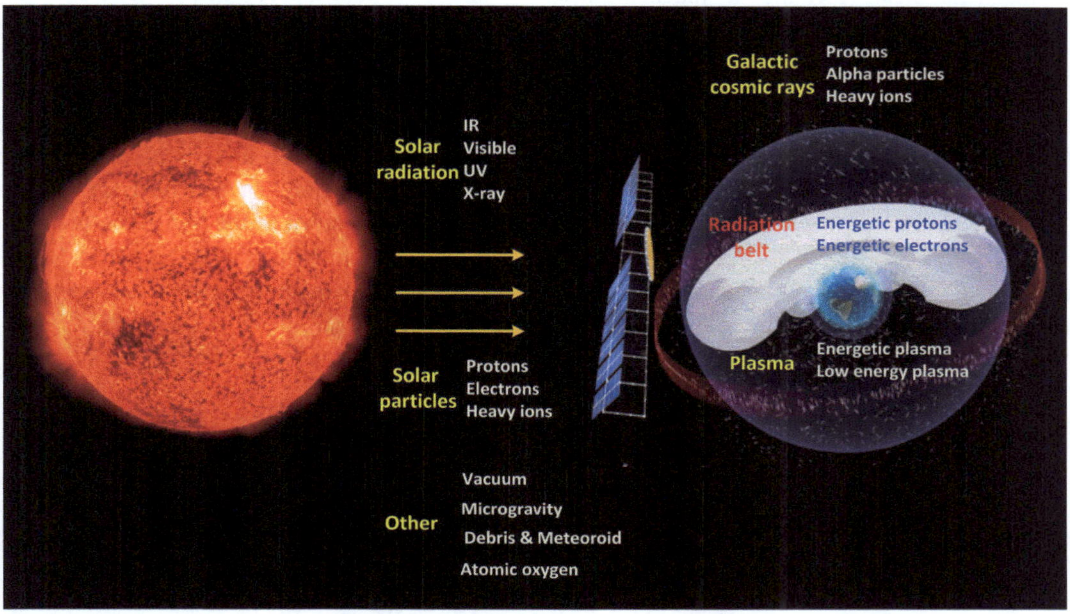

Fig. 2.5 Typical space environment factors. (Credit: Xinbin Hou)

240 nm in solar radiation. The volume density of atomic oxygen is 10^6–10^9 atoms/cm^3.

6. Space plasma environment

Under the combined action of the sun's electromagnetic radiation, particle radiation, and the earth's magnetic field and residual atmosphere of the thermosphere, an ionosphere plasma and magnetosphere plasma environment will be formed in earth's space. The ionosphere plasma region is located in the earth's space at a height of 60 km to several thousand kilometers. The molecules and atoms in the residual atmosphere of the earth's upper layer are ionized by the sun's ultraviolet rays, X rays and high-energy particles, producing free electrons and positive and negative ions, which are in a state of partial or complete ionization. On a macroscopic level, they appear quasi-electrically neutral, with low particle energy below 1 eV, but high plasma density ranging from 10^4 to 10^6/cm^3, reaching its peak near an altitude of about 300 km.

The plasma originating from the solar wind and the plasma from above the ionosphere to the magnetosphere boundary are affected by the interaction of the solar wind and the earth's magnetic field, forming a magnetosphere plasma region (Fig. 2.6). On the Sun-facing side of the earth, the magnetosphere boundary is compressed by the pressure of the solar wind, bringing it closer to the earth. On the side of the earth facing away from the sun, the solar wind stretches the magnetic field, forming a long columnar tail that can extend up to 10^6 km. The plasma environment in the magnetosphere plasma region is very complex. During solar calm periods, this region is filled with cold plasma ($E \leq 10$ eV), with a high density (>1/cm^3). During solar activity, disturbances in the solar wind and interplanetary magnetic field will cause large disturbances in the magnetosphere, generating magnetic storms and substorms, accelerating the energy of electrons and particles to above 1 keV. The accelerated high-energy charged particles will be injected into the magnetosphere under the action of the magnetic tail electric field, captured by the geomagnetic field. And the energy and density of the particles in the magnetosphere plasma will greatly increase to form a harsh space charged particle environment.

7. Space charged particle radiation environment

The space charged particle radiation environment mainly includes high-energy electrons, protons and a small number of heavy ions, mainly from the earth's radiation belt, solar cosmic rays, and galactic cosmic rays. Among them, solar cosmic rays and galactic cosmic rays are mainly extremely high-energy, low-flux protons, and heavy ions. The earth's radiation belt is due to the existence of the earth's magnetic field, and charged particles of different energies from the sun are captured by the geomagnetic field to the corresponding areas around the earth, forming a non-uniformly distributed high-intensity charged particle region, which is the most important charged particle radiation environment in earth's space, also known as the Van Allen radiation

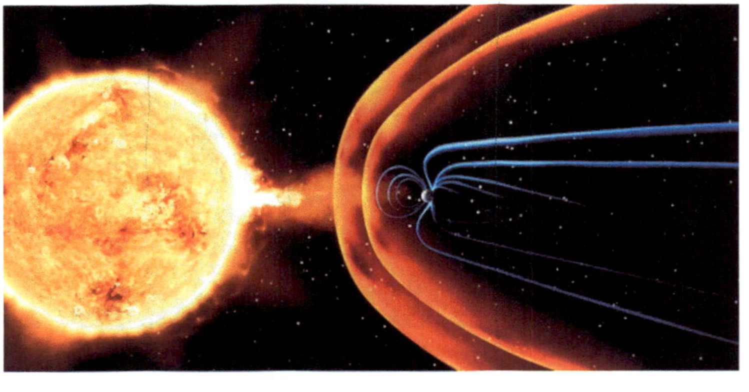

Fig. 2.6 Schematic diagram of Earth's magnetosphere

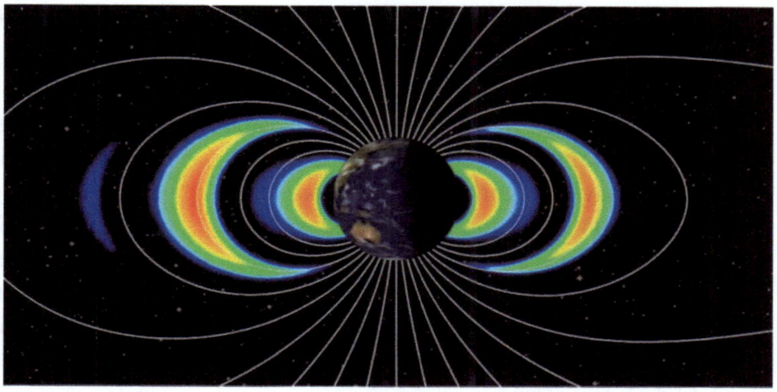

Fig. 2.7 Schematic diagram of Earth's radiation belt

belt, mainly including two space regions, namely the inner radiation belt and the outer radiation belt, as shown in Fig. 2.7.

The inner radiation belt is a charged particle region close to the earth, with an altitude range of 600–10,000 km (0.01–1.5 Re, Re is the earth's radius) on the equatorial plane, and a latitude range of about 40° north and south, mainly composed of captured protons with energies of 1–400 MeV and captured electrons with energies of several hundred thousand electron volts. The inner radiation belt is controlled by the earth's magnetic field and is relatively stable. The outer radiation belt is farther from the earth, with an altitude range of 10,000–60,000 km (1.5–10.0 Re) on the equatorial plane, the center height of the charged particle intensity is 20,000–25,000 km, and the latitude range is 55–70° north and south. The outer radiation belt is mainly captured electrons, with an energy range of 0.04–4 MeV, and also includes a small number of protons, but the energy is very low, usually in the mega electron volts. The outer radiation belt is greatly affected by the dramatically changing geomagnetic tail, with huge fluctuations in particle density, up to 1000 times.

8. Space debris and micrometeoroid environment

Space debris mainly refers to objects left in space by human space activities, including disabled satellites, final stages of launch vehicles, ejected objects, fragments produced by explosions or collisions, etc. With the increase of space activities, the number of space debris is also increasing, mainly distributed near the low Earth orbit and the geostationary orbit. Debris at lower orbits will re-enter and burn up in the atmosphere due to atmospheric drag, while debris at higher orbits will remain in orbit for a long time.

Micrometeoroids mainly come from comets and disintegrated asteroids, moving along elliptical orbits under the influence of the sun's gravitational field, with speeds relative to the Earth of 11–72 km/s, and an average speed of about 20 km/s. Comets are mainly composed of ice particles mixed with high-density mineral substances, with an average density of 0.5 g/cm^3; asteroids are mainly composed of high-density mineral substances, with an average density of 8 g/cm^3. The diameters of micrometeoroid particles in low Earth orbit are mostly between 50 μm and 1 mm.

2.2.2 The Impact of the Space Environment on SPSs

1. Vacuum environment

For SPSs, the vacuum provides a very clean environment, the surface of the solar array will not be contaminated, and power transmission will not cause loss. However, the substances produced by vacuum outgassing of materials and material decomposition in the vacuum environment can deposit on the surface of solar cells or optical devices, causing a decrease in the power generation of the solar array and the performance of optical devices.

The vacuum environment has a significant impact on heat dissipation, which can only rely on radiation heat exchange with the external environment. In addition, the adhesion and cold-welding effects caused by high vacuum can have a significant impact on space moving parts. Also, the launch process from the ground to space will experience a low vacuum environment of 10^3–10^{-1} Pa, which can easily cause low-pressure discharge in electrified equipment.

2. Cold dark environment

Due to the existence of the cold dark background in space, the heat generated by internal and external heat sources can be dissipated through the surface of the equipment by thermal radiation, thus maintaining an appropriate temperature. However, for surfaces without heat input, long-term radiation to the cold dark background environment can lead to extremely low temperatures.

3. Microgravity environment

The microgravity environment in space has a significant impact on the structural design of the station. The structure basically does not need to consider the strength problems caused by microgravity, and should mainly consider the mechanical performance of the assembly module in the launch state, the mechanical performance of the module in orbit deployment and on-orbit assembly, and the mechanical performance of the entire SPS during attitude and orbit control in space. In addition, the microgravity environment will also have a significant impact on the design of the heat dissipation fluid circuit and the storage and utilization of liquid fuel.

4. Solar electromagnetic radiation environment

The visible light and infrared radiation from the sun can be directly used for solar power generation and can also be absorbed by surface materials and converted into thermal energy. However, the ultraviolet radiation from the sun can damage the chemical bonds of materials, causing material decomposition, discoloration, reduced elasticity, and other phenomena, thereby changing the optical and mechanical properties of the materials, greatly affecting the performance and lifespan of spacecraft. In addition, due to the large area, the light pressure generated by solar electromagnetic radiation is significant, which can greatly affect the attitude and orbit control of the entire SPS.

5. Atomic oxygen environment

Atomic oxygen has high chemical activity and severe oxidation. The impact of atomic oxygen can cause material thickness loss, surface state changes, and severe erosion of solar arrays, exposed cables, and exposed organic materials, leading to a decline in material performance. During the launch and deployment process of an SPS, it needs to stay in low orbit for a long time, during which the erosion effect of atomic oxygen must be fully considered.

6. Plasma environment

The impact of the plasma environment mainly includes two aspects. On the one hand, microwave wireless power transmission interacts with ionospheric plasma. The nonlinear interaction between microwave energy and the ionosphere can cause unstable changes in ionospheric parameters, possibly causing interference with communication and navigation. At the same time, the unstable changes in the ionosphere can affect the microwave pilot beam, thereby affecting the accuracy of the power transmission beam. On the other hand, space plasma can cause charging and discharging on the surface of equipment. For GEO, under severe conditions such as geomagnetic substorms, a large amount of high-energy plasma can charge the material surface to an extremely high negative potential. Due to the different lighting conditions and material characteristics of various spacecraft surfaces, there can be a large potential difference between surfaces, leading to electrostatic discharge. Electrostatic discharge occurring in high-voltage solar arrays can potentially cause short circuits, result in the failure of the entire solar array, and pose a fatal threat. This is a key issue that an SPS needs to pay attention to.

7. Space charged particle radiation environment

High-energy charged particles in space can impinge on the interior of materials or devices, causing different effects, thereby causing different damage effects on materials or devices and destroying their performance. This mainly includes total ionizing dose effects, displacement damage effects, single particle effects, and internal charging effects.

The total ionizing dose effect mainly comes from electrons and protons in the earth's radiation belt and solar cosmic ray protons. Charged particles impinging on the interior of materials or devices cause ionization, and the energy is absorbed by the absorber. The accumulation of high-flux charged particles will cause total ionizing dose damage to components or materials, leading to performance degradation and failure of components or materials, such as changes in the performance of thermal control coatings, reduction in the strength and performance degradation of polymer insulating materials, changes in transistor performance, etc.

Displacement damage effects mainly come from high-energy protons in the earth's radiation belt and solar flare protons. After high-energy protons impinge on the interior of materials or devices, they can cause lattice atoms to displace, causing lattice defects, thereby causing displacement damage. Displacement damage has a significant impact on the semiconductor materials of solar cells, reducing the power generation efficiency of solar cells and affecting the power generation performance of SPSs.

The single particle effect mainly comes from high-energy protons in the earth's radiation belt, galactic cosmic rays, and high-energy heavy ions and protons from solar cosmic rays. Heavy ions and protons generate a large number of charged particles in the sensitive areas of electronic devices, causing soft errors, transient failures, or permanent failures in semiconductor devices. The main effects include single particle upset, single particle latchup, single particle gate rupture, single particle burnout, etc., which have a significant impact on electronic devices, especially high-power high-voltage power management devices.

The internal charging effect mainly comes from high-energy electrons in the earth's radiation belt. High-energy electrons can penetrate the surface of materials and deposit inside dielectric materials. Long-term incidence of high-energy electrons causes the charge deposition to reach a certain level, forming an electric field that may break down the dielectric material, triggering internal discharge in the material. Minor internal discharges can interfere with electronic systems, severe discharges can damage dielectric materials, reduce insulation strength, and even cause short circuits and damage to electronic devices. This has a significant impact on high-voltage power transmission cables, high-voltage conductive rotary joints, and power management devices of SPSs.

8. Space debris and micrometeoroid environment

The solar arrays, concentration systems, and transmitting antennas of Solar Power Satellite are huge, so they are highly likely to be hit by micrometeoroids and space debris, making protection difficult. Micrometeoroids and space debris hitting the solar array can cause damage to the PV cells and local circuits, and may also cause local discharge of the solar array, resulting in failure of some areas of the array. Large-area concentrating mirrors hit by micrometeoroids and space debris may cause the mirrors to tear, reducing the overall efficiency of the concentrating system. The transmitting antennas hit by micrometeoroids and space debris can cause damage and failure of local antenna units, affecting the power transmission efficiency.

2.3 Composition of SSP

2.3.1 Engineering Composition of the SSP Project

The SSP project is a mega system, and its construction, operation, and management are all very grand projects. The entire SSP project is mainly divided into six parts (Fig. 2.8).

1. Solar power satellite system
   The SPS mainly operates in the GEO, responsible for converting space solar energy into electric power, and efficiently converting electric power into microwave power, continuously transmitting power to the ground receiving system at a fixed point.
2. Transportation system
   The transportation system is mainly used to transport the modules of SPSs to the operating orbit position (GEO) and to provide necessary equipment and material supplies for the operation and maintenance of SPSs. It mainly includes ground-to-LEO launch rockets and LEO-to-GEO orbital transfer vehicles.
3. Launch site system
   The launch site system mainly meets the assembling, testing, and launching needs of launch rockets, and meets the assembling and testing needs of the modules of SPSs.
4. Space construction and support system
   The space construction and support system is mainly used to support the on-orbit assembly of modules carried out during the on-orbit construction of SPSs, and to provide necessary on-orbit maintenance and material reserves for the stable operation of SPSs. It mainly includes space assembly service platforms and assembly robots.
5. Ground receiving system (Ground application system)
   The ground receiving system is responsible for continuously receiving the energy transmitted by the SPS, converting it into electric power with the high efficiency, and connecting it to the power grid after regulation.
6. Ground operation control system (telemetry and control system)
   The ground operation control system includes two functions: first, it provides necessary telemetry and remote operation support for the launch, on-orbit construction, and operation of SPSs; on the other hand, it provides a reverse beam guidance signal for the SPS, and monitors the working status of SPSs and the ground energy receiving system to ensure the stable operation of SPSs and the ground energy receiving system.

The SPS and the ground receiving system are the core systems of the entire project, while the transportation system, launch site system, space

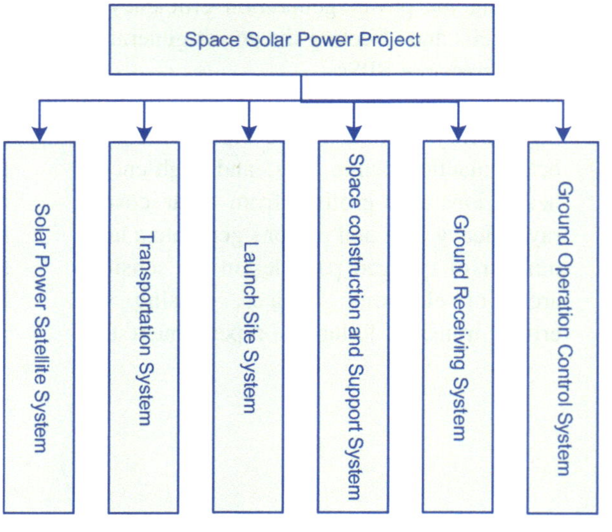

Fig. 2.8 Composition of space solar power project

construction and support system, and ground operation control system are indispensable important support systems for realizing this grand project. The specific composition of each system will be determined according to the technical approach and specific implementation plan.

2.3.2 Composition of an SPS

In order to meet the functional requirements of power generation in space and continuous power transmission to the ground, an SPS needs multiple components, mainly including space solar energy collection, space solar energy conversion, space power transmission and management, microwave wireless power transmission, laser wireless power transmission, structure, attitude and orbit control, thermal control, information and system operation management, etc. (Fig. 2.9). The functions of each component are as follows.

1. Space solar energy collection: Effectively collects solar energy and transmit it to the photovoltaic conversion device
2. Space solar energy conversion: Efficiently converts the solar energy collected in space into electric power
3. Space power transmission and management: Transmits high-power electric power to the wireless power transmission devices and supplies power to other service electronic devices.
4. Microwave wireless power transmission: Converts electric power into high-power microwave and uses large-diameter transmitting antennas to transmit microwave power to the ground.
5. Laser wireless power transmission: Converts electric power into high-power laser and uses large-diameter optical systems to transmit laser power to the target.
6. Structure: Serves as the support for the entire SPS, connects all parts together, and provides the rigidity and strength required for attitude and orbit control.
7. Attitude and orbit control: According to the SPS's requirements, realizes the Sun orientation of the solar energy collection part and the Earth-pointing of the wireless power transmission device, and maintains the SPS's orbit position and orbit altitude.
8. Thermal control: Mainly manages and dissipates the heat of all parts of the entire SPS, ensuring a reasonable temperature range for each part of the equipment.

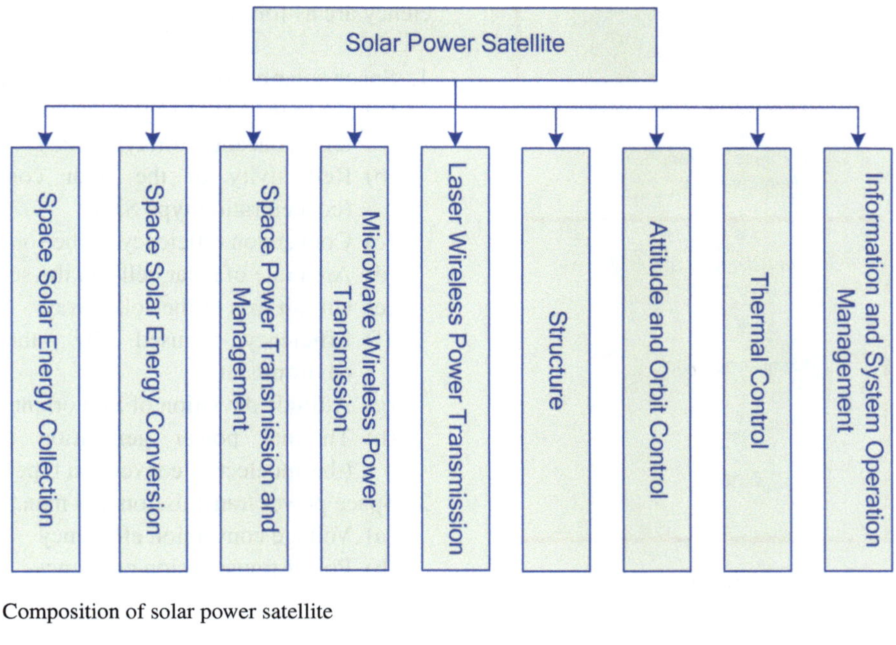

Fig. 2.9 Composition of solar power satellite

9. Information and system operation management: Collects the operating data and information of all parts of the entire SPS and carries out unified operation control and management of the entire system.

2.3.3 Energy Efficiency Chain of the SSP System

The energy conversion and transmission process of a typical SSP system mainly includes space solar energy collection and conversion, power transmission and management, energy conversion and emission, wireless power transmission, power reception and conversion, power regulation and grid connection, as shown in Fig. 2.10.

Therefore, for the SSP, the electric power received by the ground power grid is

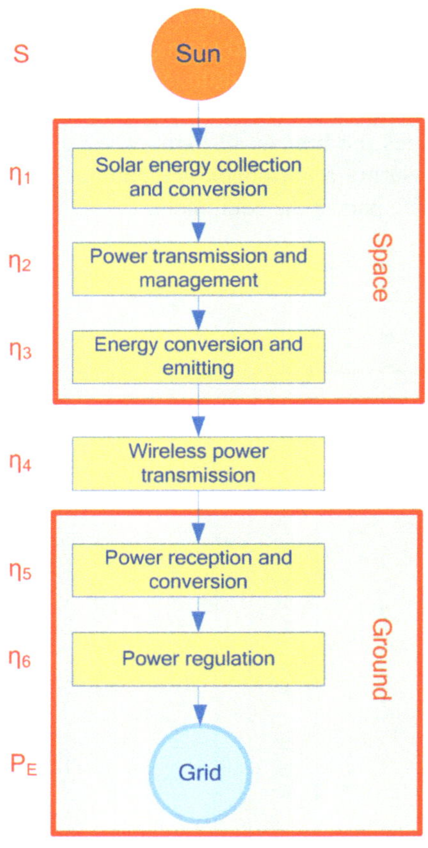

Fig. 2.10 Energy efficiency chain of SSP

$$P_E = \eta_1\eta_2\eta_3\eta_4\eta_5\eta_6 SA \tag{2.1}$$

where

P_E—Electric power input the grid on the ground
η_1—Efficiency of space solar energy collection and electric power conversion
η_2—Efficiency of space electric power transmission and management (considering transmission loss and power consumption of service equipment)
η_3—Efficiency of converting electric power into microwave or laser and energy emission
η_4—Transmission efficiency of electromagnetic radiation from space through the atmosphere to the ground receiving device (considering atmospheric loss and beam collection efficiency)
η_5—Efficiency of ground power reception and conversion (including the reception efficiency and energy conversion efficiency of the receiving device on the ground)
η_6—Efficiency of power regulation and grid integration on the ground
A—Effective collection area of space solar energy, m^2
S—Solar radiation intensity on the earth's orbit, about 1360 W/m^2

The main parameters related to system efficiency are as follows:

1. Space solar energy collection and conversion
   (a) Efficiency of the solar concentration system (concentration type SPS)
   (b) Reflectivity of the solar concentrator (concentration type SPS)
   (c) Conversion efficiency of the solar PV cell
   (d) Area rate of solar cells on the solar array
   (e) Circuit loss of the solar array
   (f) Efficiency caused by the space environment
   (g) Attitude deviation of Sun orientation
   (h) Thermal power generation efficiency (thermoelectric conversion type)
2. Space power transmission and management
   (a) Voltage conversion efficiency
   (b) Power transmission efficiency

   (c) Power regulation and distribution efficiency
3. Energy conversion and emission
   (a) Frequency (wavelength)
   (b) Size of the emitting antenna (optical system)
   (c) Electric-microwave (electric-laser) conversion efficiency
   (d) Beam direction control precision
   (e) Beam synthesis efficiency
   (f) Emission device efficiency
4. Wireless power transmission
   (a) Atmospheric loss of power transmission
   (b) Space loss of power transmission (beam collection efficiency, directly related to the beam direction precision, size of the emitting antenna, and size of the reception device)
5. Ground power reception and conversion
   (a) Size of the reception device
   (b) Energy absorption efficiency
   (c) Energy conversion efficiency
   (d) Power conversion, collection, and transmission efficiency

2.4 Comparison of Various Wireless Power Transmission Methods

SSP needs to use wireless power transmission technology to transmit energy from space to the ground, with a transmission distance of about 36,000 km, and the transmission process needs to pass through the earth's atmosphere. It is necessary to consider the scale of energy emission device and reception device and the loss through the atmosphere (Fig. 2.11). The long-distance wireless power transmission methods currently considered mainly include three types: direct reflection of sunlight, microwave wireless power transmission (MPT), and laser wireless power transmission (LPT). The comparison of the three methods is shown in Fig. 2.12.

The direct reflection of sunlight refers to the deployment of large-area solar concentrators in space, reflecting sunlight directly from space to the ground. The advantage of the direct reflection of sunlight is that it does not require multiple energy conversion processes, and the solar array on the ground can directly convert space solar energy into electric power. The biggest problem with the direct reflection of sunlight, apart from the great influence of weather on the reflection of sunlight, is that the sun is a sphere light source (with an angle of view of 32″). Therefore, it is limited by geometric optics. The sunlight incident after reflection by the concentrator actually forms an image of the sun at the receiving end. According to the diameter of the sun, the distance between the sun and the earth, and the distance of wireless power transmission, it can be concluded that the ideal optical reflection situation on the ground has a light spot diameter of about 330 km, and the size of the concentrator only determines the intensity of the reflected light. Obviously, the direct reflection of sunlight requires a huge size of space concentrator and ground solar array, which is not practical to implement in engineering.

The microwave wireless power transmission method and the laser wireless power transmission method are similar, both first convert solar energy into electric power, then convert electric power into electromagnetic waves, and then use the emission device to transmit the electromagnetic waves to the ground, and then reconvert the electromagnetic waves into electric power. The main differ-

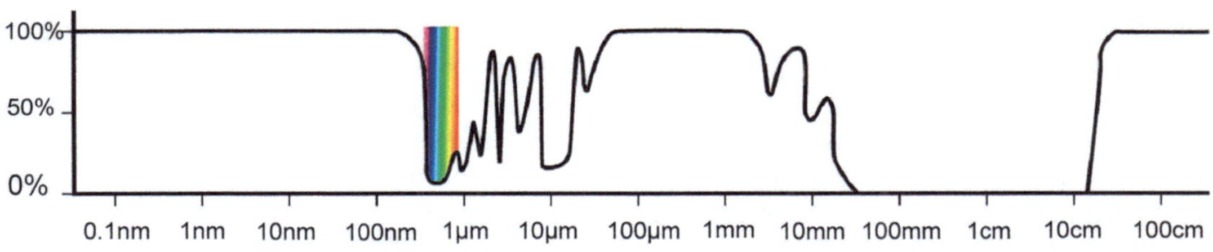

Fig. 2.11 Atmospheric transmittance of electromagnetic waves

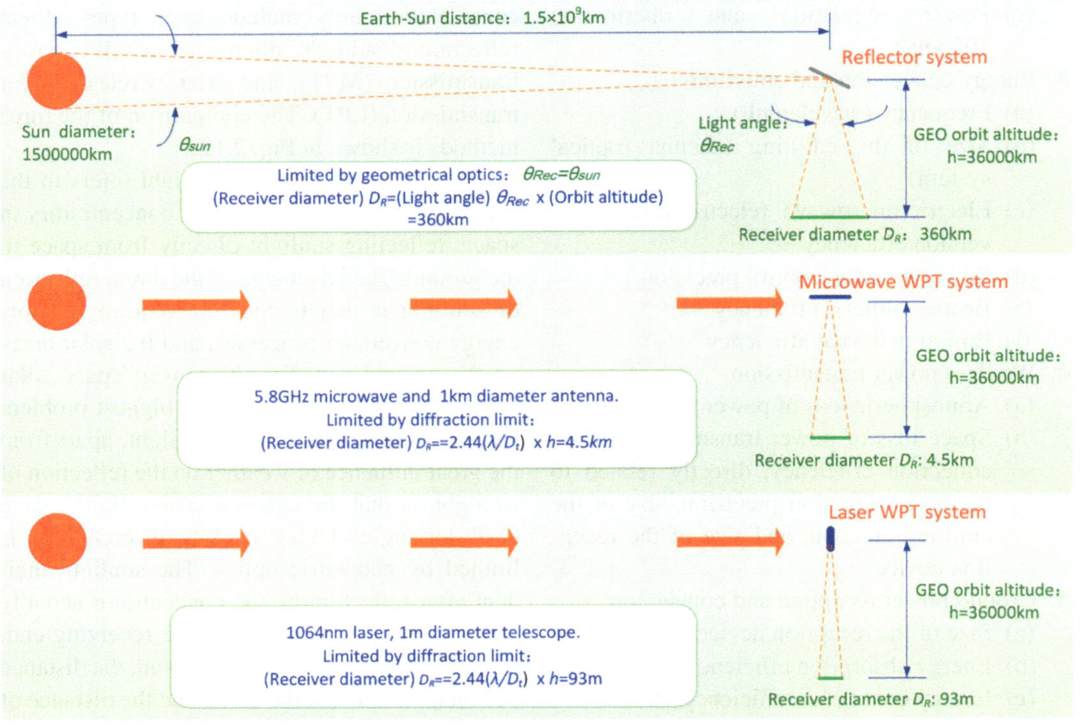

Fig. 2.12 Comparison of different wireless power transmission methods. (Credit: Xinbin Hou)

ence between the microwave method and the laser method lies in the huge difference in wavelength. According to the principle of electromagnetic wave transmission, the product of the emitter diameter and the receiver diameter is proportional to the wavelength. Assuming that the microwave uses 5.8 GHz, the corresponding wavelength is 5 mm, and the laser uses 1064 nm, the difference between the two is nearly 5000 times, so the corresponding device scale is also huge difference. For microwave power transmission, the typical sizes of the emitting antenna and the receiving antenna are 1 and 4.5 km. For laser power transmission, the typical sizes of the transmission optical system and receiving laser PV array are 1 and 93 m (under ideal optical conditions).

Microwave wireless power transmission and laser wireless power transmission are the two transmission methods currently considered in the SSP scheme. The comparison between microwave power transmission and laser power transmission is shown in Table 2.1. The laser transmission beam is narrow, the corresponding transmission and receiving devices are small, the direction control is flexible, and it is suitable for medium and low power wireless power transmission in space. The advantage of microwave power transmission is that the impact of atmosphere and rain on transmission is small, the conversion efficiencies of the emitting antenna and receiver are high, the power density is low, the safety is good, and it is suitable for high-power transmission from space to the ground. For power transmission from space to the ground, the loss through the atmosphere must also be considered. In order to transmit microwaves more efficiently in the atmosphere, the Industrial, Scientific and Medical (ISM) band, which is basically not affected by weather conditions such as clouds and rain, is generally considered, mainly using 2.45 or 5.8 GHz microwave frequencies. Laser power transmission mainly uses visible light or near-infrared spectrum in the atmospheric transparent windows and also needs to consider factors such as energy conversion efficiency and beam quality to determine the frequency.

Table 2.1 Comparison of microwave power transmission and laser power transmission

Factors	Microwave power transmission	Laser power transmission
Frequency	Low	High
Beam angle	Wide	Narrow
Transmitter size	Large	Small
Receiver size	Large	Small
Loss through atmospheric	Small	Middle
Affected by weather	Small, can penetrate clouds	Large
DC-RF efficiency	High	Lower
RF-DC efficiency	High	Medium
Thermal control	Low heat flux density, easy thermal control	High heat flux density, difficult thermal control
Receiving device	Rectenna, only used for microwave conversion	Photovoltaic technology, can also be used for solar light conversion
Pointing accuracy requirement	Higher	Extremely high
Pointing control	Mechanical pointing + beam phase control	Mechanical pointing
Transmission interference	Interference with communication and electronic devices	Interference with astronomical observations
Safety	Beam power intensity is low, relatively safe	Need to limit laser power intensity below sunlight
Military use	Rare	Can be used for military purposes, need to limit power intensity
Transmission method	Suitable for centralized wireless power transmission	Can achieve distributed wireless power transmission

Bibliography

1. Hou X B, Zhang X H, Wang L. Introduction on Space Solar Power[M] (in Chinese). Beijing: China Astronautic Publishing House, 2020.
2. Wang L, Guo S L, Xu N J, et al. Introduction on Spacecraft Radiation Hardening Technology[M] (in Chinese). Beijing: China Astronautic Publishing House, 2021.

3 Typical Solar Power Satellites

3.1 Classification of SPSs

In order to ensure the efficient operation of the SPS, it is necessary for the solar array (or concentrator) to be oriented towards the sun and the transmitting antenna to be oriented towards the receiving device on the ground. For GEO, the relative position change between the solar array (or concentrator) and the transmitting antenna is 360° in one orbit period. Due to the large volume, mass, and power of the SPS, how to handle the mutual position relationship between the solar array and the transmitting antenna becomes a core issue to consider in the design of the SPS. The current SPS schemes generally consider the following situations:

1. Use high-power conductive rotary joints to achieve relative rotation between the solar array and the transmitting antenna
2. Use a concentration scheme to reflect sunlight to the solar array through the control of the concentrator system, without the need for high-power conductive rotary joints
3. Use microwave reflection method to change the direction of microwave through the rotation of large microwave reflectors, without the need for high-power conductive rotary joints
4. Not use a rotating mechanism, the relative position of the solar array and the transmitting antenna is fixed, at the cost of discontinuous power generation and system efficiency loss

At present, dozens of SPS concepts have been proposed internationally, including several new concepts proposed in recent years, mainly based on microwave power transmission, but none of the schemes is optimal. By comparing various SPS concepts, they can generally be divided into three types (Fig. 3.1), namely non-concentration continuous transmission type SPS, discontinuous transmission type SPS, and concentration type SPS.

3.2 Non-concentration Continuous Transmission SPS

3.2.1 1979 SPS Reference System

Starting in 1977, based on Glaser's SPS concept, the US Department of Energy (DOE) and the National Aeronautics and Space Administration (NASA) collaborated with Boeing Co. and Rockwell Co. to implement the Satellite Power System Concept Development and Evaluation Program (CDEP). By 1980, approximately \$50 million had been invested in the research and evaluation of the SPS system, resulting in the famous "1979 SPS Reference System" with a power generation capacity of 5 GW (Fig. 3.2).

The design of this system places the SPS in a geostationary orbit, with the overall structure consisting of a solar array and a microwave

X. Hou, *Space Solar Power*, Sustainable Development Goals Series,
https://doi.org/10.1007/978-981-97-3580-8_3

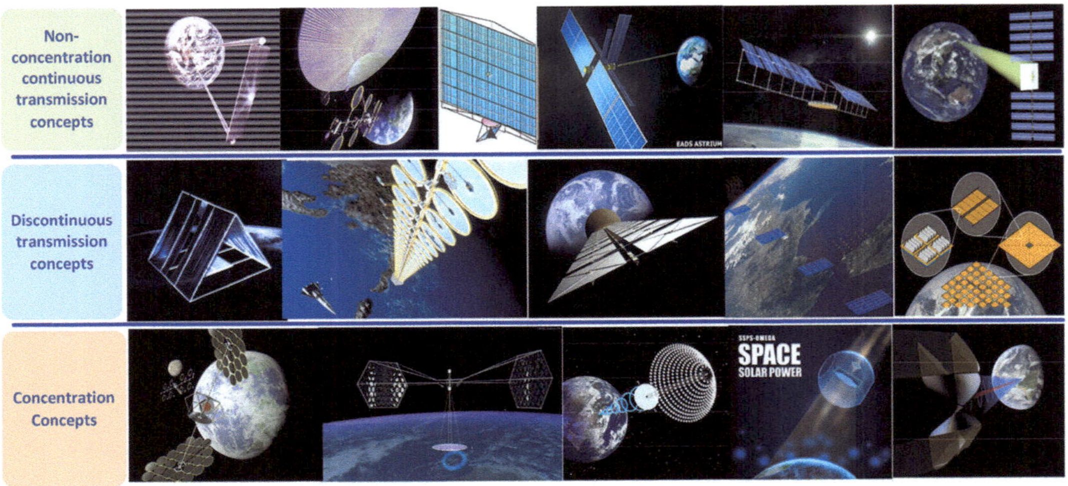

Fig. 3.1 Classification of typical SPSs. (Credit: Xinbin Hou)

Fig. 3.2 Illustration of 1979 SPS reference system (Solar Power Satellite and Ground Receiving Antenna). (Credit: NASA)

transmitting antenna. The two components are connected by a large conductive rotary mechanism to achieve solar array sun tracking and ground-pointing of the microwave transmitting antenna (Fig. 3.3). The diameter of the microwave transmitting antenna is 1 km, with a microwave frequency of 2.45 GHz. The corresponding ground rectenna has a diameter of about 10 km (for equatorial region), and for the location at a latitude of 35°, the antenna is a 10 km × 13 km elliptical area. The power generation on the ground is 5 GW, with the overall system power generation efficiency at about 7%. The solar array is a massive structure with a total area of 55 km^2. The biggest technical challenge of this design is the large conductive rotary mechanism, which needs to transmit nearly 9 GW of electric power from the solar array to the microwave transmitting antenna. Although the total transmission power is high, the power intensity of the microwave beam reaching the ground is relatively small, with the beam center at about 230 W/m^2 and the edge only at 10 W/m^2. The main design parameters of the system are shown in Table 3.1, with a total mass of 30,000–50,000 tons.

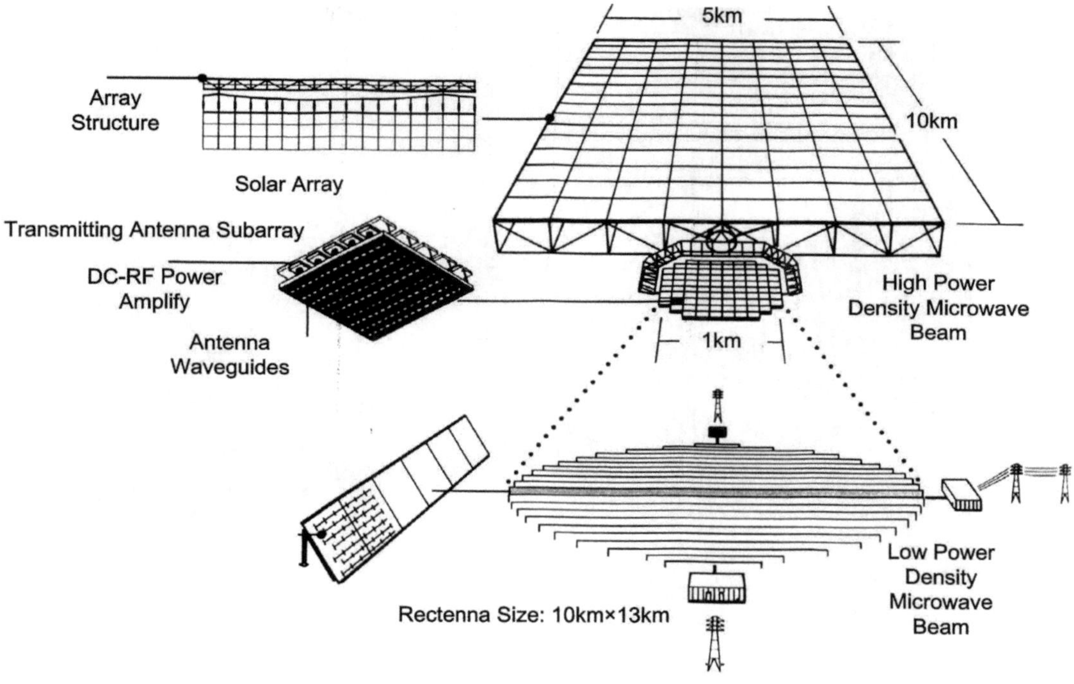

Fig. 3.3 1979 SPS reference system scheme. (Credit: NASA)

Table 3.1 Parameters of 1979 SPS reference system

Design parameters		Value
System parameters	Orbit	Geostationary orbit
	Power generation/GW	5
	Life/years	≥30
	Mass/t	3×10^4 to 5×10^4
Solar array	PV cell	Silicon or gallium arsenide
	Size/km	$10 \times 5 \times 0.5$
	Material	Carbon fiber composite material
Power transmission	Microwave conversion	Speed control tube
	Transmitting antenna diameter/km	1
	Frequency/GHz	2.45
	Ground receiving antenna size/km	10×13 (ellipse)
	Maximum power intensity/(W/m^2)	230 in the center, 10 at the edge

3.2.2 Multi-Rotary Joint Solar Power Satellite

To solve the problem of ultra-high-power conductive rotary joint, China Academy of Space Technology proposed the Multi-Rotary joint Solar Power Satellite (MR-SPS) in 2014. This concept designs a special configuration (Fig. 3.4), which decomposes the whole solar array into multiple independently rotating solar subarrays. Each solar subarray transmits electricity through an independent conductive rotary joint. Therefore, the technical problem of the 1979 SPS reference system's high-power conductive rotary joint is solved by using multiple rotary joints, and the single point failure problem of the conductive rotary joint is avoided. At the same time, the solar array, truss structure, and microwave transmitting antenna of the entire SPS are all modularly designed, which is also conducive to the assembly and construction of the system.

Fig. 3.4 Illustration of MR-SPS. (Credit: Xinbin Hou)

The MR-SPS mainly consists of three parts: solar array (south, north), microwave transmitting antenna, main structure. The solar array and the microwave transmitting antenna are connected through the main structure. Service subsystem equipment is installed on the structural framework of the solar array, microwave transmitting antenna, and main structure. The solar array is composed of multiple solar subarrays, each of which generates and transmits electric power to the cable of the main structure through two conductive rotary joints, and then transmits the power to the microwave transmitting antenna over a long distance. The main structure consists of two north-south main truss structures and multiple up-down main truss structures. The north-south main truss structures are used to support the solar subarray and the microwave transmitting antenna, and the multiple conductive rotary joints installed on the upper main truss structure are used to drive the solar subarray to achieve its sun-oriented. The two north-south main truss structures are connected together through the up-down main truss structures, which are used to transmit power to the microwave transmitting antenna through the cable. The entire SPS adopts a modular design and is built in geostationary orbit using an on-orbit robot assembly method. For more details, see Chap. 7.

3.2.3 K-SSPS

South Korea proposed the K-SSPS concept in 2019 (Fig. 3.5). This concept is very similar to MR-SPS, but does not use multiple rotary joints configuration. The center of the entire SPS is a square transmitting antenna with a side length of 1 km, the microwave frequency is 5.8 GHz, and the two sides are connected to the huge solar array in the north and south through conductive rotary joints. The solar array uses CIGS or perovskite flexible solar cell, including 4000 roll-out solar array modules, with a total area of 11.2 km^2. The power generated by the solar array is transmitted to the microwave transmitting antenna through the cable and conductive rotary joints. The diameter of the ground receiving antenna is about 4 km, the entire SPS is designed to generate power of 2 GW, the system efficiency is 13.5%, and the main parameters are shown in Table 3.2. The entire system needs to be assembled in low Earth orbit, and

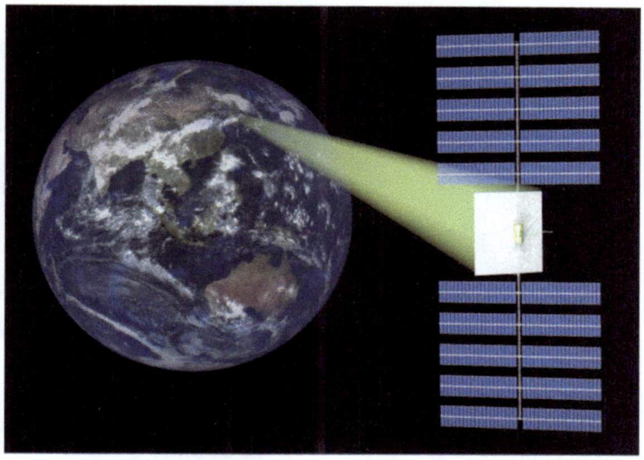

Fig. 3.5 Illustration of K-SSPS concept. (Credit: KARI)

Table 3.2 Parameters of K-SSPS

Parameter	Value
Ground power supply/GW	2
External dimension/km	2 × 5.6
Solar Array (CIGS or Perovskite flexible solar cells)	Roll-out thin-film
Total system efficiency/%	13.5
Solar array size/km	2 × 2.7 × 2 units
Solar array module size/m	10 × 270
Number of solar array modules	4000
Transmitting antenna area/km^2	1.0
Operational orbit	Geostationary orbit
Microwave frequency/GHz	5.8
Receiving antenna diameter/km	4

then use the electric propulsion system to transport to geostationary orbit. In order to prevent the solar PV cells from being affected by the radiation environment during the orbit transfer process, only some of the solar array modules are unfolded and work normally to supply power to the high-power electric propulsion system (Fig. 3.6). After reaching the geostationary orbit, all solar array modules are unfolded, and the entire system begins to operate normally.

3.2.4 Modular Multi-Rotary Joint Solar Power Satellite

In 2021, China Academy of Space Technology proposed an improved concept based on the MR-SPS concept, named the Modular Multi-Rotary joint Solar Power Satellite (MMR-SPS) concept, as shown in Fig. 3.7. Its main technical feature is: the circular microwave transmitting antenna array originally located in the center of the entire SPS is transformed into a rectangular microwave transmitting antenna array. This array is composed of multiple rectangular transmitting antenna subarrays, arranged on the truss structure under the solar subarray, realizing the one-to-one correspondence between the solar subarray and the microwave transmitting antenna subarray. The power generated by the solar subarray is directly transmitted to the corresponding microwave transmitting antenna subarray by the corresponding conductive rotary joint. This design greatly simplifies the difficulty of space power transmission and management, and each module is completely independent during the assembly process of the entire SPS, which improves the reliability of the system. Moreover, each module can generate electricity independently after assembly, which improves the utilization efficiency of the SPS.

The MMR-SPS can be divided into three major parts when viewed from the earth's direction: the solar array, the main structure, and the microwave transmitting antenna. From the north-south direction, it includes multiple power generation and power transmission modules composed of solar subarrays, connecting truss structures, and microwave transmitting antenna subarrays. These modules extend in the north-south direction to form the entire SPS. For a typical 1 GW system, the entire SPS includes 50 power generation and power transmission modules (Fig. 3.8). Each solar subarray measures

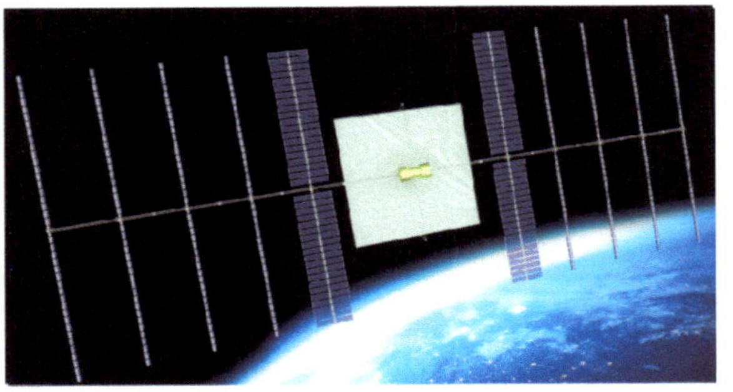

Fig. 3.6 K-SSPS in low Earth orbit. (Credit: KARI)

Fig. 3.7 Illustration of MMR-SPS. (Credit: Xinbin Hou)

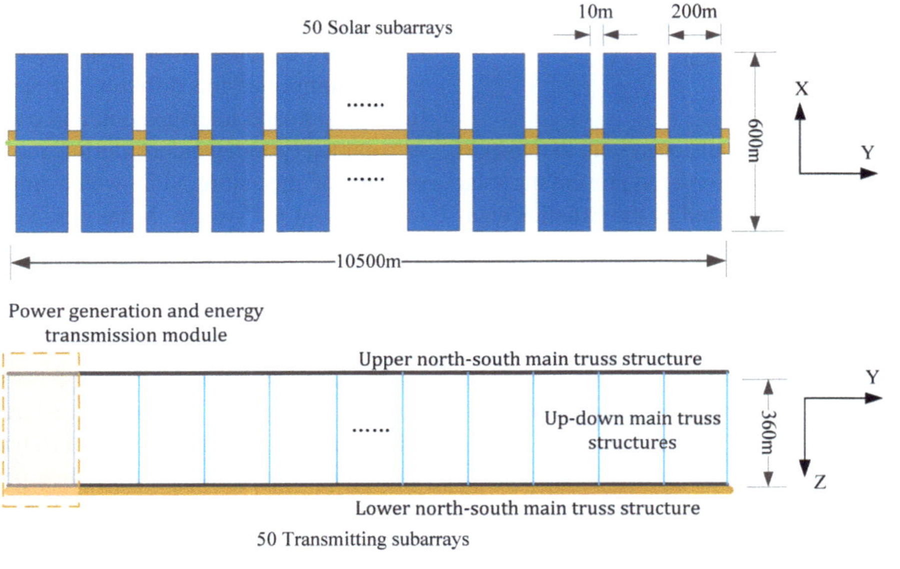

Fig. 3.8 Configuration of MMR-SPS

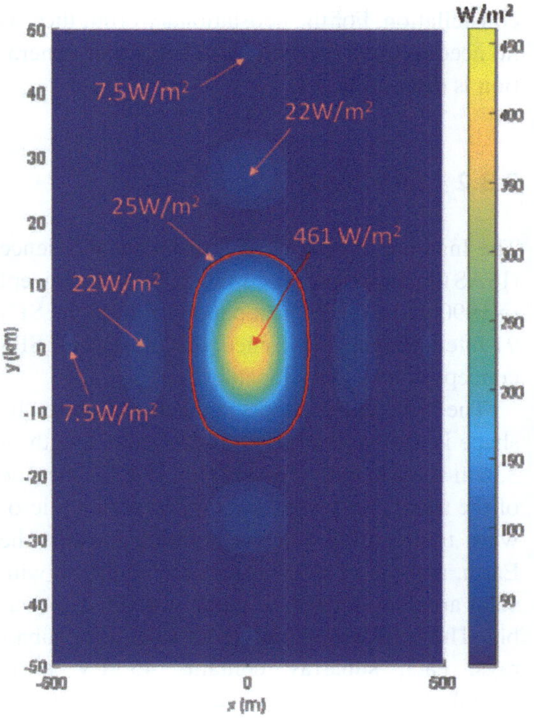

Fig. 3.9 Microwave beam distribution on the ground

200 m × 600 m and is connected to the microwave transmitting antenna subarray below through the longitudinal connection of the main truss structure. Each microwave transmitting antenna subarray measures 210 m × 100 m. The final size of the microwave transmitting antenna is 10,500 m × 100 m, forming a long and narrow beam shape on the ground (Fig. 3.9).

3.3 Discontinuous Transmission SPS

3.3.1 SunTower SPS

In the mid-1990s, NASA restarted a new study of SSP called "Fresh Look" from 1995 to 1997. Based on the evaluation of dozens of new SSP concepts, two concepts were selected, one of which was the SunTower SPS.

The SunTower SPS consists of multiple solar array modules, a main frame, and a microwave transmitting antenna. The solar array of the

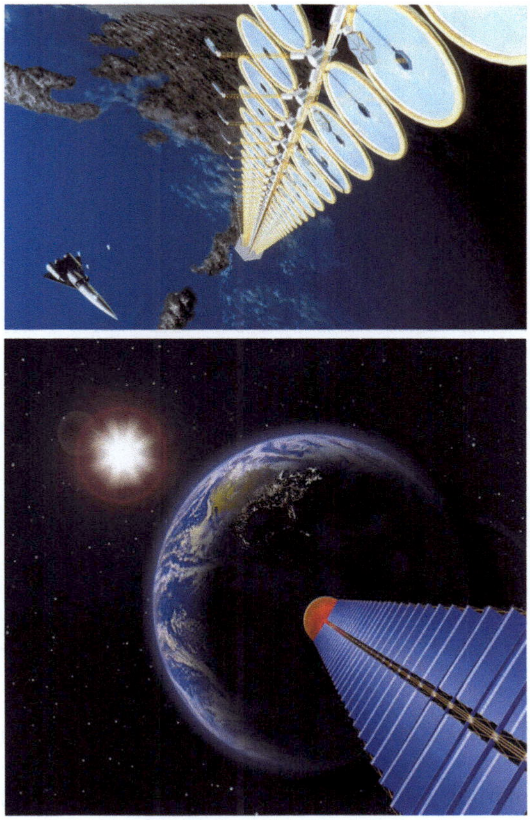

Fig. 3.10 Illustration of SunTower SPS (Concentrator Solar Array and Non-Concentrator Solar Array). (Credit: NASA)

SunTower SPS consists of dozens to hundreds of solar array modules, which can adopt inflatable concentrator solar array structures or non-concentrator solar array structures (Fig. 3.10). A typical solar array module has an output power of about 1 MW, and the number of solar array modules is configured according to the total power generation demand. The solar array modules are installed in two columns on the central backbone, or multiple frames can be used, with two columns of solar array modules installed on each backbone (Fig. 3.11). The generated power is transmitted to the microwave transmitting antenna at the end of the backbone through the central superconducting cable installed on the main backbone. There is no relative rotation between the solar array and the microwave transmitting antenna. Its biggest technical feature is the use of gravity gradient stabilization, which makes the

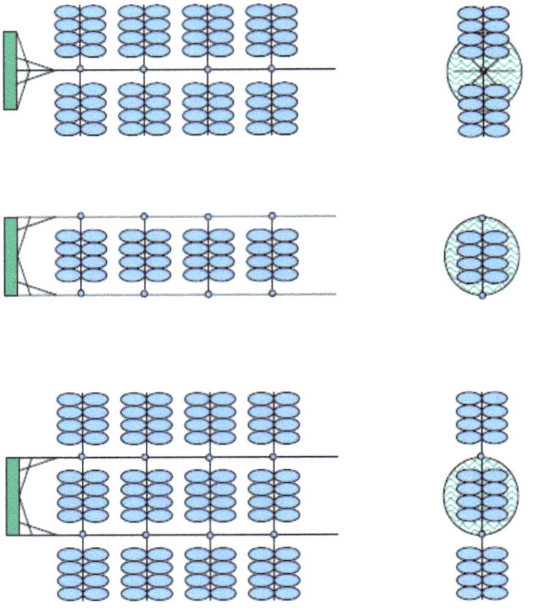

Fig. 3.11 Multi-backbones SunTower SPS structure. (Credit: NASA)

central structure automatically perpendicular to the ground, ensuring that the microwave transmitting antenna at the end is aimed at the ground, greatly reducing the difficulty of attitude and orbit control. However, because it is impossible to achieve solar arrays Sun-oriented, power generation is discontinuous.

The SunTower SPS can operate in sun-synchronous orbit, equatorial orbit, or geostationary orbit. For the sun-synchronous orbit, the solar array can maintain a good Sun-oriented direction, and it needs to be composed of multiple SPSs to form a constellation, and multiple ground receiving stations are also needed. For the equatorial orbit, it is impossible to achieve solar arrays Sun-oriented, so power generation is discontinuous and needs to be composed of multiple SPSs to form a constellation. For example, for a scheme operating at a height of 12,000 km in the equatorial orbit, the microwave frequency is selected as 5.8 GHz, the diameter of the transmitting antenna is about 260 m, the diameter of the ground receiving antenna is about 4 km. The beam control ability of the transmitting antenna is ±15°, it can cover a range of 30° north and south latitude, and six SPSs are needed to form a constellation. For the geostationary orbit, there is no need to use a constellation, but power generation is discontinuous.

3.3.2 SPS 2000

The Institute of Space and Astronautical Science (ISAS) in Japan proposed the SPS 2000 concept in 1990 and won the best paper award at the SPS 91 International Conference. It is the earliest SPS concept proposed by Japan.

The SPS 2000 adopts a fixed structure. Its shape is a triangular prism, with a side length of 336 m and a height of 303 m (Fig. 3.12). One side of the triangular prism is equipped with a microwave transmitting antenna pointing towards the Earth, and the other two sides are equipped with solar arrays to receive as much sunlight as possible. The solar array consists of four solar subarrays, each subarray contains 45 PV array modules, each module serves as the basic structure unit for assembly, with an output voltage of 1 kV, current of 180 A, and a mass of 270 kg. The microwave transmitting antenna is a 132 m × 132 m square phased array antenna, equipped with 1936 square subarrays. The side length of the subarray is 3 m, which is the basic unit of phase control, and each subarray includes 1320 antenna units. The selected microwave frequency is 2.45 GHz, the beam scanning angle range is longitudinally ±30°, laterally ±16.7°, and the microwave transmission power is 10 MW. The diameter of the ground receiving antenna is 2 km, and the average power generation of each receiving station is about 300 kW. The system parameters of SPS 2000 are shown in Table 3.3.

As a demonstration SPS, considering the reduction of launch costs and the reduction of the distance for power transmission from space, SPS 2000 is designed to operate in a 1100 km equatorial orbit. In order to achieve as continuous energy reception as possible at the ground receiving station, multiple SPSs need to form a constellation. It is planned to demonstrate power supply by supplying power to developing countries along the equator.

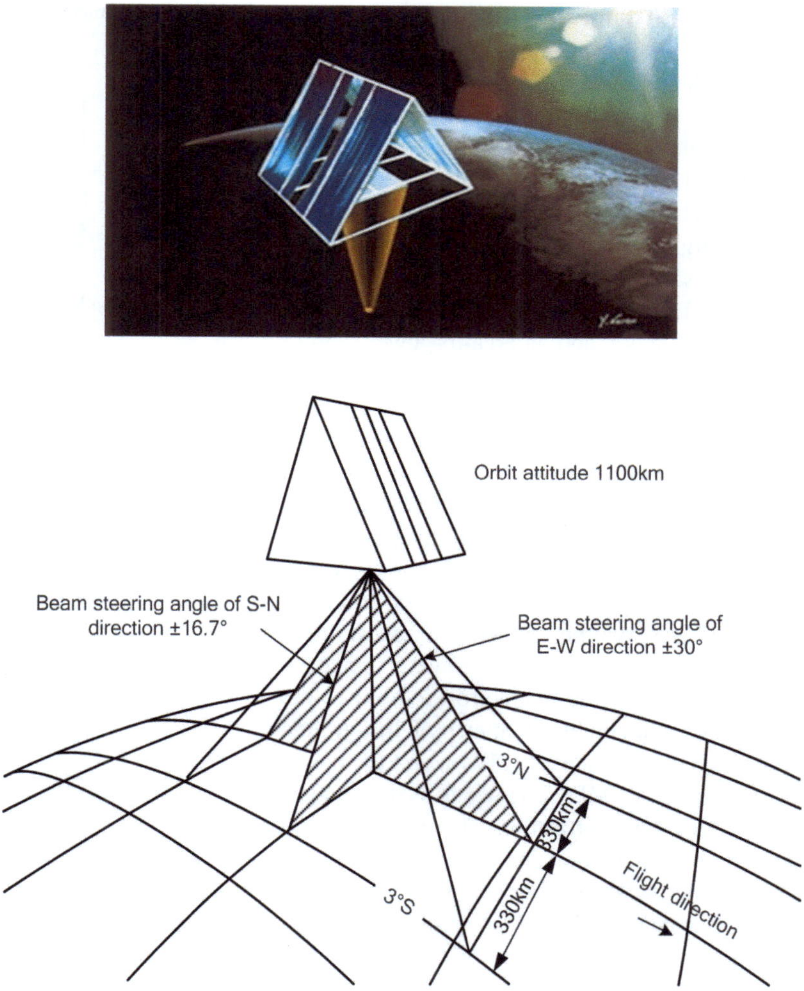

Fig. 3.12 Schematic diagram of SPS 2000. (Credit: ISAS)

Table 3.3 Parameters of SPS 2000

Parameter	Value
Configuration	Triangular prism
Size	336 m × 303 m
Mass	134.4 t
Operating orbit	Equatorial circular orbit, altitude 1100 km
Microwave frequency	2.45 GHz
Beam control method	Retrodirective beam control
Beam scanning angle	−30° to 30° (East-West)
	−16.7° to 16.7° (North-South)
Antenna size	132 m × 132 m
Number of antenna units	2,547,776
Microwave transmission power	10 MW
Receiving antenna diameter	2 km
Ground receiving power	300 kW
Type of solar cell	Thin-film amorphous silicon cell
Voltage of solar array bus	1 kV
Power generation	32 MW
Attitude control method	Gravity gradient stabilization

3.3.3 Sail Tower

From 1998 to 1999, ESA proposed the concept of Sail Tower (Fig. 3.13) concept in the research of Space Exploration and Utilization (SE&U). This concept is based on the American SunTower SPS concept. The main feature is that the solar array adopts a lightweight thin-film structure similar to a solar sail. Each solar array module is 150 m × 150 m in size, with a mass-to-power ratio of 225 W/kg, and an area-to-power ratio of 170 W/m^2. There are 60 pairs, 120 solar sail array modules symmetrically installed on a 15 km long central structure, and the maximum on-orbit power generation is about 450 MW. The central structure is equipped with superconducting cables, which are used to transmit the power generated by the solar array to the microwave transmitting antenna at the end of the central structure. The diameter of the microwave transmitting antenna is about 1 km, the microwave frequency is 2.45 GHz, and the magnetron is used as the microwave source; the diameter of the ground receiving antenna is about 11 km, and the power generation is 275 MW. The whole system adopts gravity gradient stabilization. There is no relative motion between the microwave transmitting antenna and the solar array, so the power generation is discontinuous. Each solar sail array module automatically unfolds after being launched into orbit, the system is assembled in low Earth orbit, and then sent to geostationary orbit by the electric thrusters. The size of the overall system is 15 km × 0.35 km × 0.05 km, with a total mass of about 2140 t. The main parameters of Sail Tower concept are shown in Table 3.4.

Fig. 3.13 Illustration of Sail tower. (Credit: ESA)

Table 3.4 Parameters of the sail tower

Parameter		Value
Orbit		GEO
SunTower	Length/km	15
	Mass/t	2140
	Maximum on-orbit power generation/MW	450
Solar array	Number of modules	120
	Module size/m	150 × 150
	Module weight/t	2.5
Central structure and cable	Weight/t	240
	Cable	Superconducting cables
Transmitting antenna	Number of magnetrons	400,000
	Frequency/GHz	2.45
	Diameter/m	1020
	Weight/t	1600
	Transmitting power/MW	400
Rectenna	Antenna size/km	11 × 14
	Size including the safety area/km	27 × 30
	Power generation/MW	275

3.3.4 Tethered-SPS

In 2000, Japan proposed a new concept, Tethered-SPS, which is currently the basic type of microwave wireless power transmission SPS that Japan focuses on developing. The Tethered-SPS adopts a sandwich structure design, that is, the solar cells, power management equipment, microwave conversion devices, and transmitting antennas are integrated together to form a layered structure. The platform is connected to the sandwich structure by using tethers, achieving gravity gradient attitude stability and maintaining the sandwich structure antenna facing the Earth. In the initial design, the entire SPS only has one platform, which is connected to the sandwich structure as a

Fig. 3.14 Illustration of tethered-SPS. (Credit: Japan Space Systems)

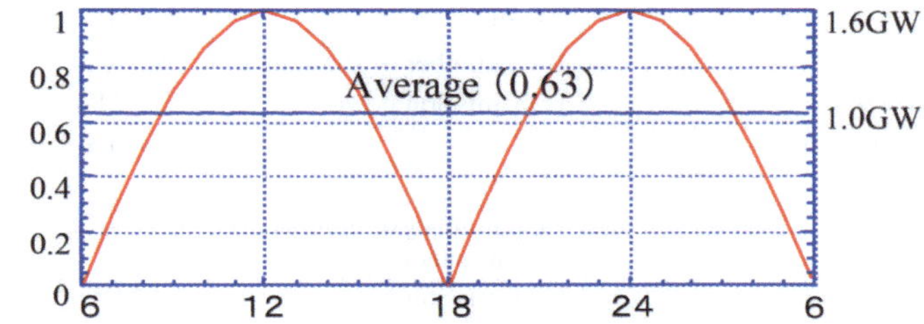

Fig. 3.15 Power generation change under the condition of double-sided solar cells

whole through four sets of tethers. In the improved design, the sandwich structure is modularized, and multiple tether structures are combined (Fig. 3.14).

The main technical features of this concept include: no need for conductive rotary mechanism; no use of concentration system, simple thermal control; using sandwich structure design, simple power management and distribution; using gravity gradient stability, simplifying attitude control. However, this concept cannot achieve Sun-oriented of solar array within the orbit period, so the power generation fluctuates greatly and shows periodic changes. Even considering the deployment of solar cells on both sides of the sandwich structure, the system utilization rate can only reach to about 63% (Fig. 3.15).

The sandwich structure size of a GW-level SPS is 2.5 km × 2.375 km, and the tether length is 5–10 km. The basic structural unit is a sandwich structure unit board with dimensions of 100 m × 95 m, each sandwich structure unit board contains 3800 antenna modules, the total mass is about 42.5 t, and the microwave power transmission power is 2.1 MW. 25 basic structural units (5 × 5) are assembled to form a sandwich structure sub-board, the size reaches 500 m × 475 m, the microwave transmission power is 52.5 MW. Each sub-board corresponds to a platform, and the sub-board and platform are connected together with four tethers to form a basic structure. 25 basic structures (5 × 5) form a complete SPS. The main technical parameters of the system are shown in Table 3.5. The relatively simple structural design of the Tethered-SPS makes it technically feasible, the highly modular design is conducive to system assembly (Fig. 3.16), and the system's expansion capability is greatly improved.

3.3.5 Microwave Sworm SPS

In 2015, the California Institute of Technology conducted research on SSP in terms of new ultra-light structures, efficient power generation, and wireless power transmission, and proposed a new

Table 3.5 Main parameters of a Tethered-SPS

Parameter		Value
System structure	Configuration	Tethered structure, sandwich structure panel suspended by 100 tethers (4 per sub-panel)
	Size	2.5 km × 2.375 km × 0.02 m
	Tether length	5–10 km
Orbit		GEO
Ground output power		1 GW
System mass	Total mass	26,600 t
	Sandwich panel mass	25,200 t
	Platform (including tethers) mass	1400 t
Microwave frequency		5.8 GHz
Efficiency	Solar cell	35%
	Microwave source	85%
Sub-panel	Size	500 m × 475 m × 0.02 m
	Weight	1010 t
	Total number of sub-panels	25(5 × 5)
	Transmission power	52.5 MW
Unit panel	Size	100 m × 95 m × 0.02 m
	Weight	40 t
	Total number of unit panels per sub-panel	25(5 × 5)
	Transmission power	2.1 MW
Module	Size	5 m × 0.5 m × 0.02 m
	Weight	10.625 kg
	Number of modules per unit panel	3800(20 × 190)
	Power generation	1181 W
	Transmission power	555 W

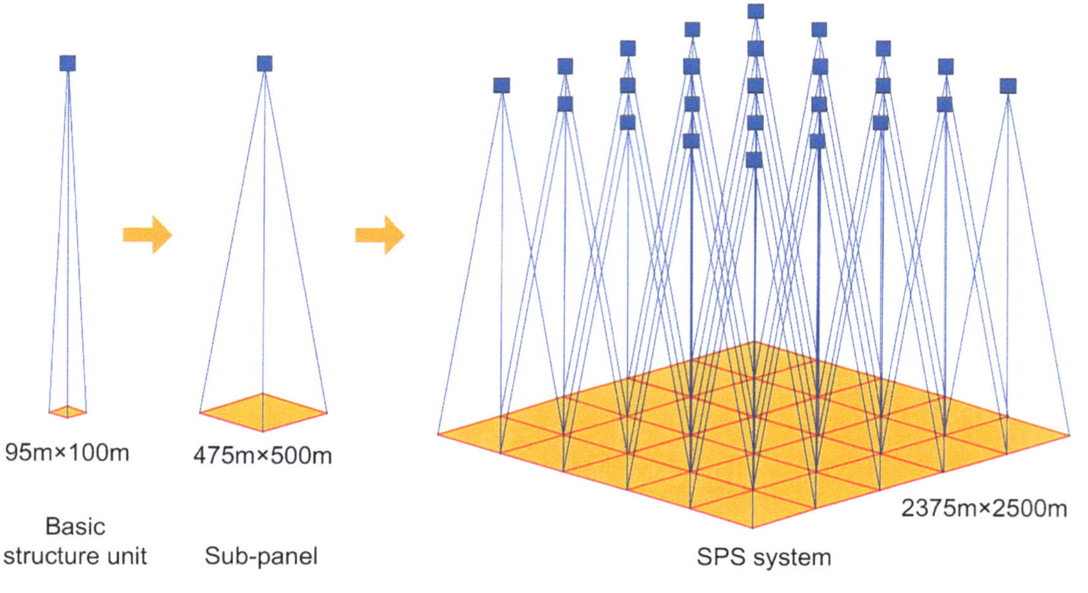

Fig. 3.16 Assembly scheme of Tethered-SPS

concept of "Microwave Sworm" SPS. In order to achieve an ultra-light and easy-to-fold structure, the research team designed a thin-film sandwich structure, including flexible thin-film solar arrays, ultra-thin CMOS RF chips, thin-film flexible antennas, and ultra-light materials and unfolding structures, using a folding and unfolding method similar to a solar sail.

Its basic unit is called a tile (Tile), composed of thin-film high-efficiency solar cells and thin-film transmitting antennas. Multiple tile units form a 1.5 m wide flexible strip component, and multiple strip components are arranged in the pattern of a solar sail to form a 60 m × 60 m basic spacecraft unit, containing 300,000 tile units. A 3 km × 3 km GW-level SPS is finally formed by 2500 basic spacecraft units (Fig. 3.17). The initial design of the thin-film high-efficiency solar array is very unique, including multiple sets of parabolic thin-film concentrators and thin-film solar cell strips matching the concentrators. The thin-film solar cell strips are installed on the edge of one side of the thin-film concentrator (Fig. 3.18). Before deployment, the concentrator and thin-film solar cell strips are pressed onto the substrate and rolled up with the flexible strip

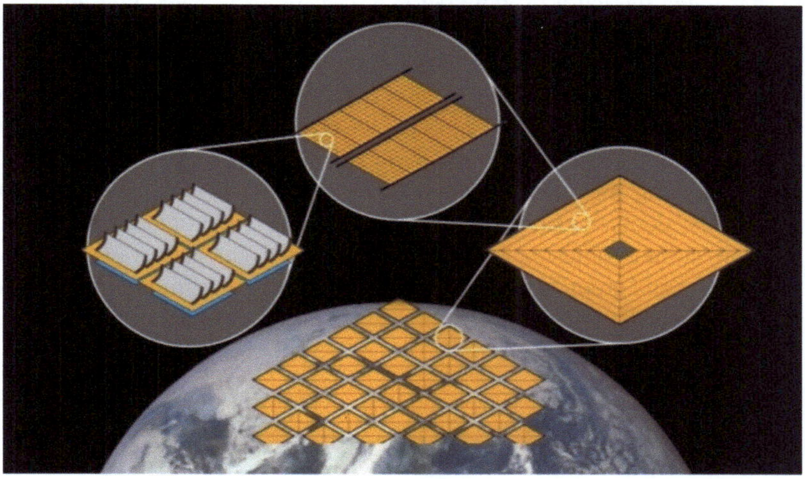

Fig. 3.17 Microwave Sworm SPS concept. (Credit: Caltech)

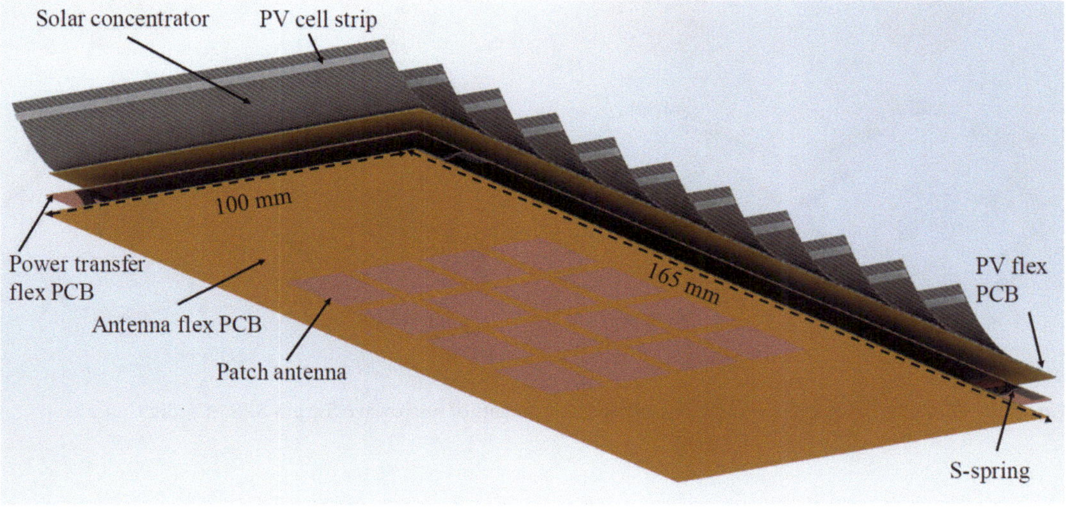

Fig. 3.18 Components of Tile unit. (Credit: Caltech)

component. After entering orbit, they autonomously unfold and recover the required shape with the deployment of the flexible strip component. The current prototype test believes that there are some problems with this concentrator design, and the latest design considers replacing the thin-film concentrator solar array with a flat solar array structure. The tile unit thin-film substrate is equipped with microwave circuits and signal circuits, etc., and the underside of the substrate is a transmitting antenna. The initial design goal of the entire spacecraft unit is to achieve an extremely low area density of 100 g/m^2, corresponding to the mass of the entire SPS of 900 t, and the latest design goal is a mass of about 2600 t.

The concept of the Microwave Sworm SPS cannot simultaneously achieve solar cell surface Sun orientation and Earth orientation of the transmitting antenna surface. The research team considers adopting measures such as double-sided solar cells, double-sided transmitting antennas, on-orbit attitude adjustment, large-angle beam adjustment, and ground energy storage system to achieve relatively stable power supply (Fig. 3.19).

3.4 Concentration SPS

3.4.1 Integrated Symmetrical Concentrator SPS

In 1999, NASA conducted the SSP Exploration Research and Technology (SERT) program, with a total investment of 22 million US dollars. The Marshall Space Flight Center (MSFC) proposed the Integrated Symmetrical Concentrator (ISC) Concept (Fig. 3.20), aiming to achieve Sun orientation of the concentrator system and Earth-orientation of the transmitting antenna through the adjustment of the concentrator system instead of using high-power conductive rotary joints. This concept mainly includes two solar concentrators, two solar arrays, and one transmitting antenna. The solar concentrators, solar arrays, and transmitting antennas are connected by the trusses. The two solar concentrators rotate as a whole to achieve Sun orientation, reflecting sunlight onto the solar array. The solar array and the antenna array are relatively fixed, powered by the transmitting antenna through a short-distance power transmission cable, while the transmitting

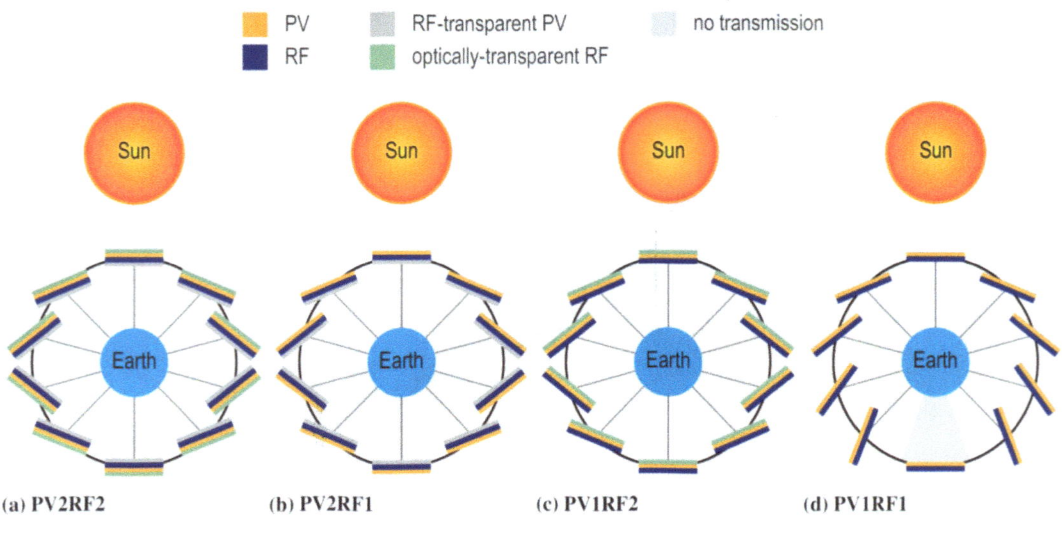

Fig. 3.19 Different sandwich structures and attitude adjustments of microwave Sworm SPS. (Credit: Caltech)

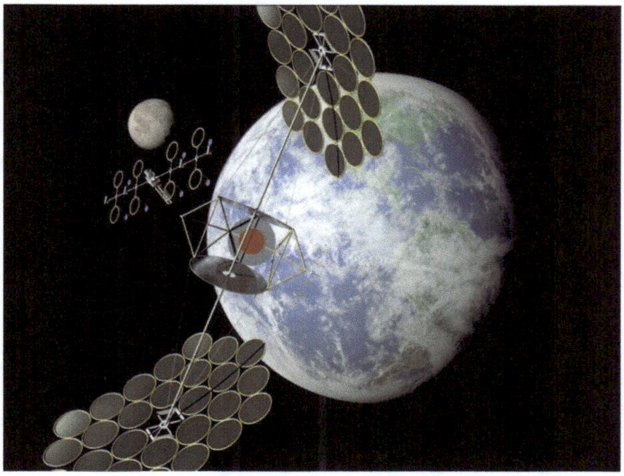

Fig. 3.20 Illustration of ISC SPS. (Credit: NASA)

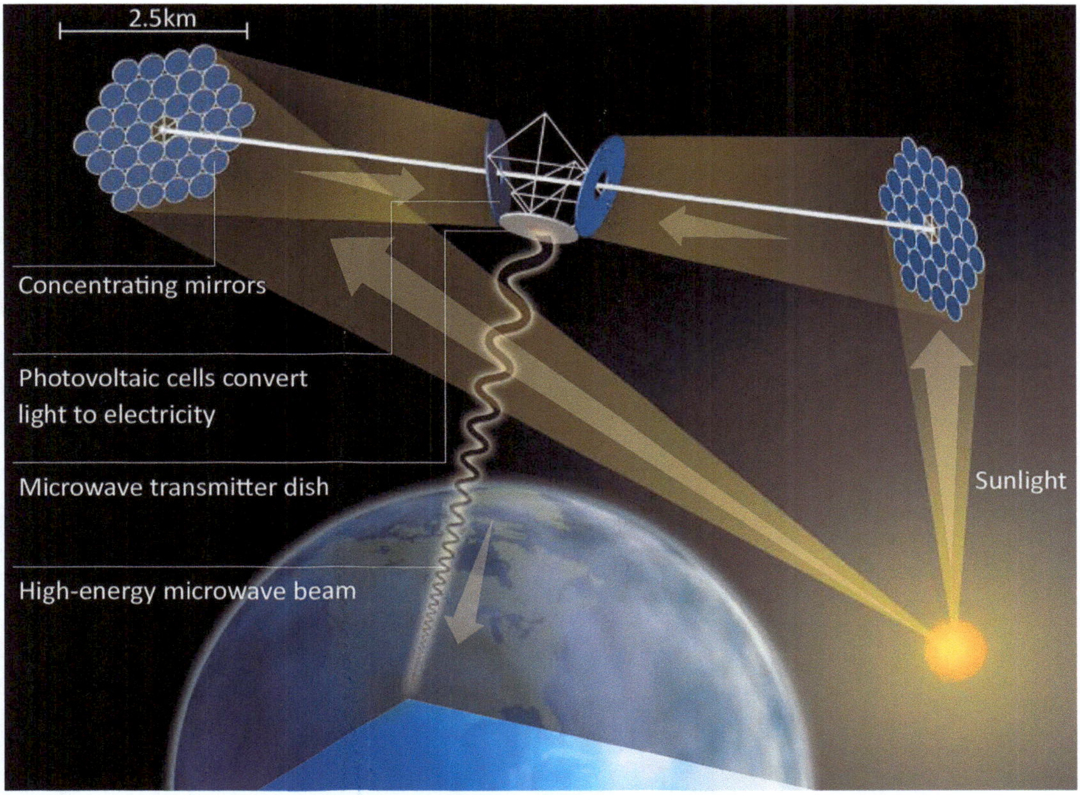

Fig. 3.21 Composition of ISC SPS. (Credit: NewScientist)

antenna maintains ground-orientation, performing microwave power transmission (Fig. 3.21).

Two large solar concentrators are symmetrically arranged to form a clamshell structure, each with dimensions of approximately 3500 m × 2500 m, a focal length exceeding 10 km, and composed of 36 reflector modules. Each reflector module is a flat mirror with a diameter of about 500 m. The mirror surface is 0.5 mm thick Kapton material, installed on the main structure of the concentrator using an inflatable structure, aiming to reflect sunlight to the solar array as much as possible, and ensuring a certain uniformity of concentration. The angle between the two solar arrays is 10°, located near the focal point of the concentrator in the center of the structure, with a diameter of 1070–1770 m, an average concentration rate of about 4.25. The expected solar cell efficiency will reach 39%, the corresponding technical indicators of solar arrays are 1 kW/kg and 550 W/m^2. The microwave transmitting antenna is arranged under the solar array and is powered by a high-voltage supply through the connection interface with the solar array. The diameter of the transmitting antenna is about 1 km, using 2.45 GHz or 5.8 GHz microwave, the DC-RF conversion efficiency is expected to reach 80%. The two solar concentrators and the solar array are connected by a truss structure with a length of about 10 km, and the total mass of the system is about 22,500 t. The ground receiving efficiency is about 90%, the RF-DC conversion efficiency of the receiving antenna is expected to reach 85%, and the final power generation is about 1.2 GW.

3.4.2 Two-Stage Reflection Modular Sandwich Microwave SPS

In 2007, the National Security Space Office of the US Department of Defense organized a study of SSP and proposed the Two-Stage Reflection Modular Sandwich Microwave SPS. The SPS includes two main reflectors, two secondary reflectors, and a sandwich panel (Fig. 3.22). The main reflector is similar to the concentrator of the ISC configuration and is composed of multiple

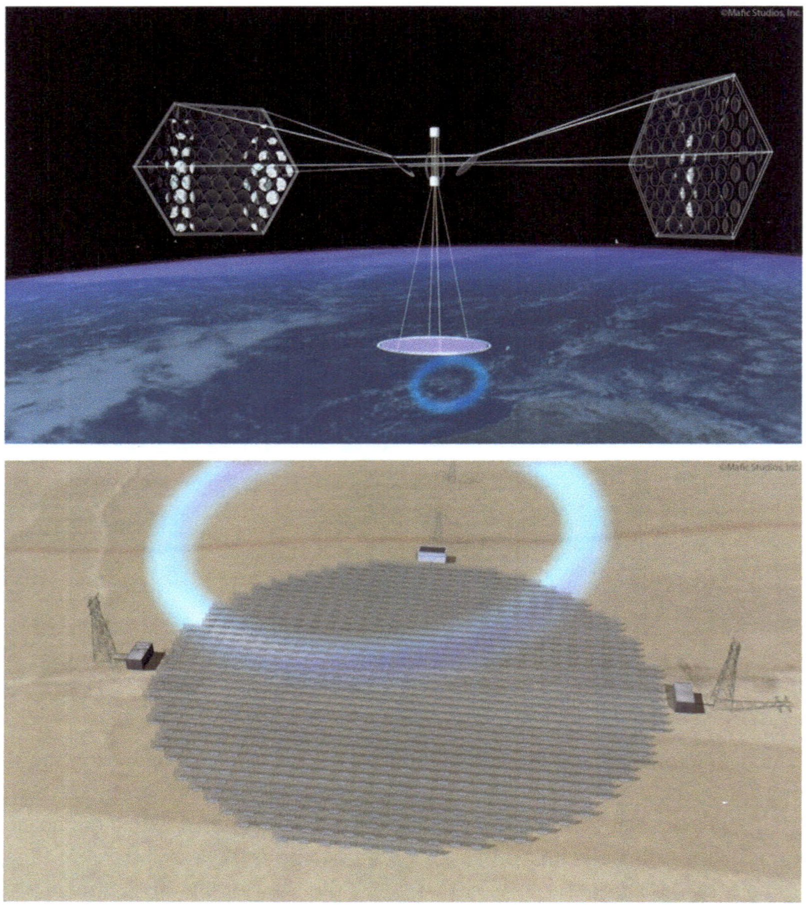

Fig. 3.22 Illustration of two-stage reflection modular sandwich microwave SPS and receiving antenna. (Credit: Mafic Studios, Inc.)

inflatable unfolding concentrator modules. The secondary reflector is a plane mirror, located on both sides near the focal point of the main reflector. The sandwich panel integrates the solar array and the transmitting antenna. The solar array is located on the upper surface of the sandwich panel, receiving the sunlight reflected by the secondary reflector, and the transmitting antenna is located on the lower surface of the sandwich panel and maintains Earth orientation. The main reflector and the secondary reflector are connected by a truss. The concentrator system and the sandwich structure are connected by a truss. The main reflectors track the sun by rotating within the orbit period and concentrate the incident sunlight on the plane of the secondary reflectors that reflect the sunlight to the solar array. For this scheme, how to control the concentrator system to ensure a relatively uniform sunlight distribution on the surface of the solar array within the entire orbit position and effective heat dissipation of the sandwich panel under high-concentration ratio are very challenging.

3.4.3 SPS-ALPHA

In 2011, with the support of NASA's Innovative Advanced Concept (NIAC) Project, John C. Mankins proposed a new SPS concept, named Solar Power Satellite via Arbitrarily Large Phased Array (SPS-ALPHA). The SPS-ALPHA is a highly modularized concentration SPS, which uses a special concentrator system. The concentrator is connected to the sandwich panel through a truss structure, forming gravity gradient stability. Its core idea is to form a huge, non-adjustable concentrator system with multiple reflector modules similar to solar sails, realizing the concentration of sunlight on the solar array on the surface of the sandwich panel. Due to its complex concentrator system, the design includes various possible configurations (Fig. 3.23), which still need to be optimized based on the concentration effect. During the Sun tracking process, each mirror module can be individually adjusted to change the reflection direction of sunlight, allowing sunlight to directly hit or reflect onto the solar

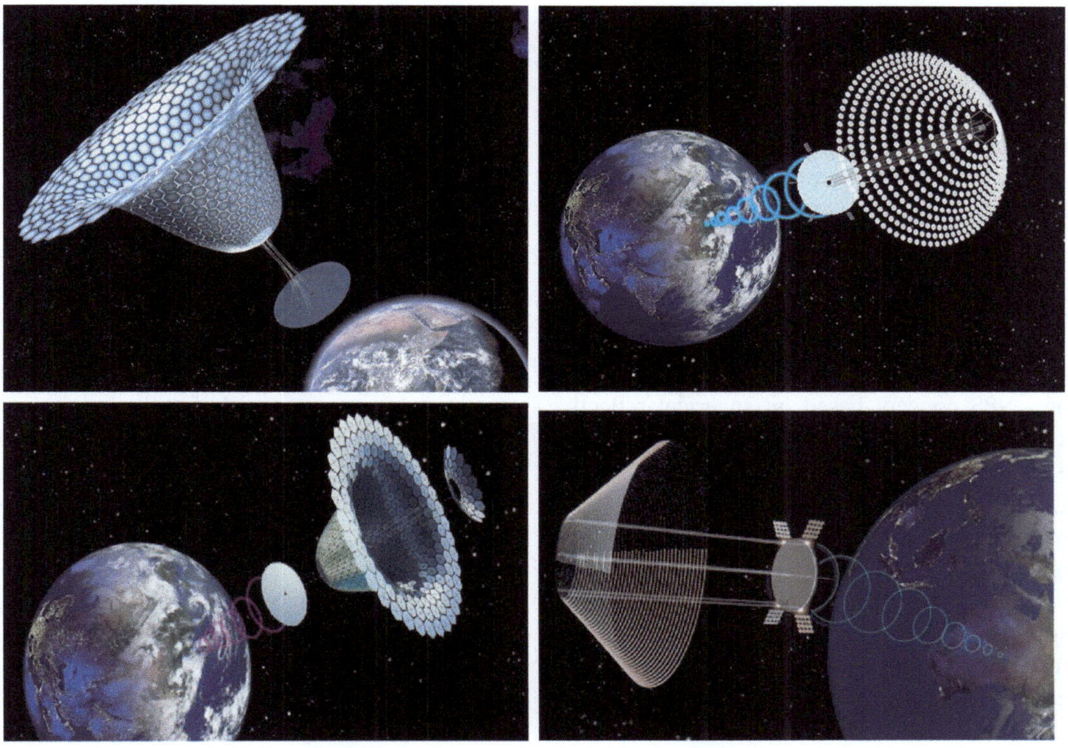

Fig. 3.23 Illustration of different SPS-ALPHA versions. (Credit: John C. Mankins)

array. The scale of the entire concentrator system will reach about 5 km.

The SPS-ALPHA design includes various typical modules: HexBus Module, Interconnects Module, HexFrame Structural Module, Reflectors and Deployment Module, Solar Power Generation (SPG) Module, Wireless Power Transmission (WPT) Module, Modular Push-Me/Pull-You Robotics (MPPR) Arm, and Propulsion/Attitude Control Module (Table 3.6). For the sandwich structure, the solar cells are located on the upper surface facing the reflector side, and the lower surface is the transmitting antenna, pointing to the ground. The basic unit of the sandwich panel is the hexagonal module, which is composed of the HexBus Module, SPG Module, and WPT Module. The hexagonal modules are connected using the Interconnects Module, eventually forming a sandwich structure with a diameter of up to 1 km. Some Propulsion/Attitude Control Modules are deployed on the sandwich panel for the orbital control of the entire system. The concentrator system and the sandwich panel are connected through the main truss structure, and their positions are fixed relative to each other. The entire system uses multiple Robotics Arms to achieve coordinated assembly between modules. Figure 3.24 shows the proposed development roadmap for SPS-ALPHA, which gradually transitions from laboratory prototypes to commercial applications.

3.4.4 SSPS-OMEGA

In 2014, Xidian University in China proposed a new SPS concept based on spherical reflective surface concentration, SSPS-OMEGA (Space Solar Power Station via Orb-shape Membrane Energy Gathering Array) (Fig. 3.25). This system mainly includes a spherical concentrating system, a solar array, and a microwave transmitting antenna array. The main idea is to use the spherical concentrating system that does not require adjustment to reduce the control difficulty of the system. The concentrating system is a spherical structure assembled from multiple thin-film planar reflector modules, removing the polar regions that cannot concentrate. The initial design is to use one-way optical thin-film materials, that is, sunlight can be incident from the front and reflected from the back. When sunlight is incident from any direction, a part of the sunlight will pass through the concentrating mirror and reflect to the surface of the solar array through the inner surface of the spherical concentrating system. In order to reduce the concentration ratio and improve the uniformity of sunlight, the solar array is designed as a conical structure and rotates at a uniform angular speed relative to the concentrating mirror and the transmitting antenna array along the spherical concentrating system, rotating once a day. The electric power generated by the solar array is transmitted to the microwave transmitting antenna through cables and conductive slide rails (Fig. 3.26). In an improved scheme, the annular slide rail is changed to two conductive rotary joints on the upper and lower sides of the antenna (Fig. 3.27). The microwave frequency is selected to be 5.8 GHz. The microwave transmitting antenna is a planar array with diameter of 1.2 km. It is located in the center of the sphere and the antenna array is connected to the north and south of the concentrating system through six cables for antenna adjustment to keep the antenna oriented to the ground.

3.4.5 CASSIOPeiA SPS

In 2017, Ian Cash from the UK proposed a new concept—CASSIOPeiA (Constant Aperture, Solid State, Integrated, Orbital Phased Array) SPS. This concept is designed as an irregular curved structure, composed of multiple layers of structures similar to the double helix structure of biological DNA, each layer includes multiple local concentrating solar arrays and special transmitting antennas that can control the beam in 360° direction (Fig. 3.28). Among them, the concentrating solar array uses a local concentrating system based on Fresnel lenses and Kohler optics, which not only reduces the quantity of solar cells, but also improves the PV conversion efficiency. With four-junction gallium arsenide cell technology, it can achieve a conversion efficiency of

Table 3.6 Main modules of SPS-ALPHA

Module element	Description	Image	Single mass/kg
HexBus module	Is the main structure of the sandwich (including wireless communication, etc.)		About 25
Interconnects module	Connects the HexBus modules together		About 1
HexFrame structural module	Provides the base structure for the reflectors and connect the reflector array to the power/transmitter array		About 50
Reflectors and deployment module	Thin-film reflectors redirect incoming sunlight to the SPG	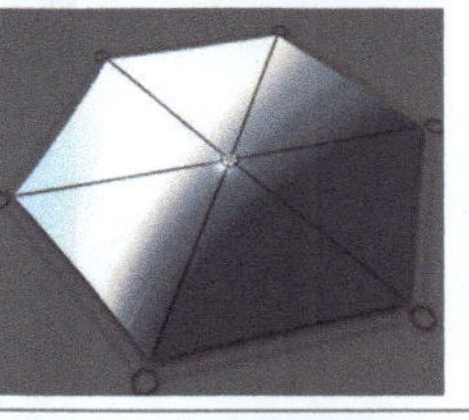	75–100
Solar Power Generation (SPG) module	Generates the power for the WPT transmitter		15–20
Wireless Power Transmission (WPT) module	Converts the electricity into a coherent RF (microwave) transmission to the receiver on Earth	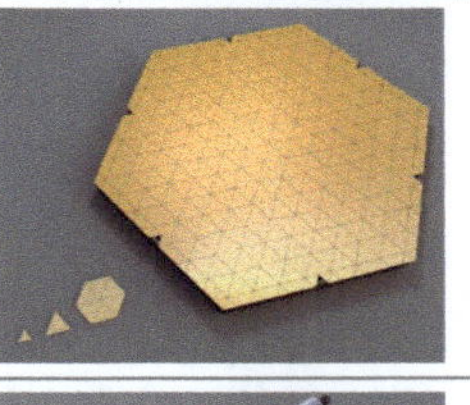	About 50
Modular Push-Me/Pull-You Robotics (MPPR) arms	Provides all sorts of In-Space Assembly and Construction	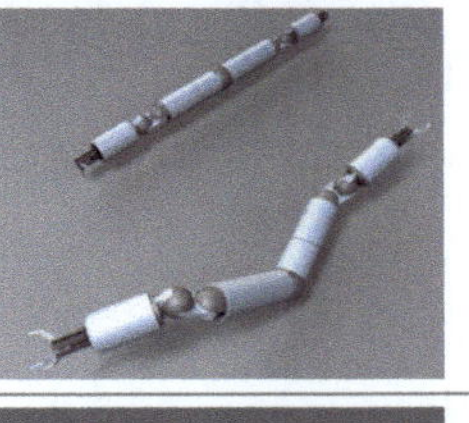	About 10
Propulsion/attitude control module	Provides the required propulsion for guidance, navigation, and control (GN&C) and station keeping for the platform	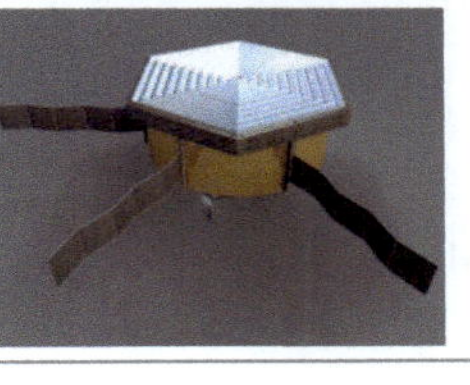	50–500

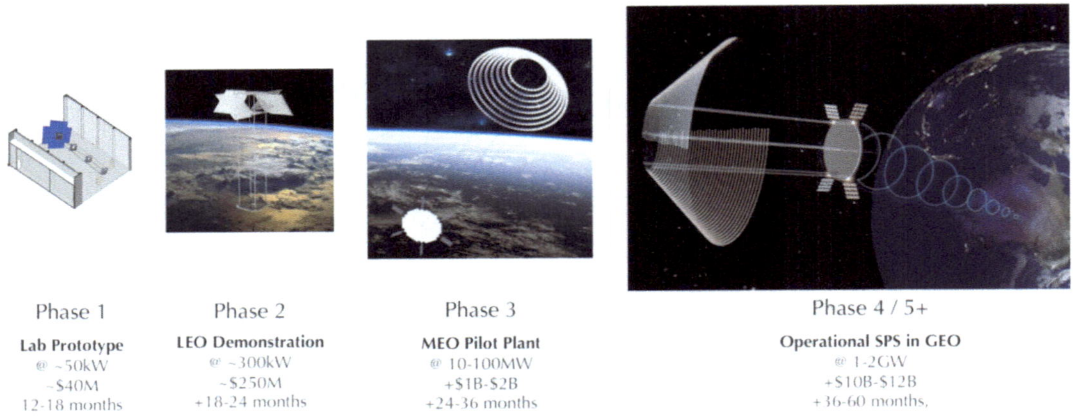

Fig. 3.24 Development roadmap of SPS-ALPHA. (Credit: John C. Mankins)

Fig. 3.25 Illustration of SSPS-OMEGA. (Credit: Xidian University)

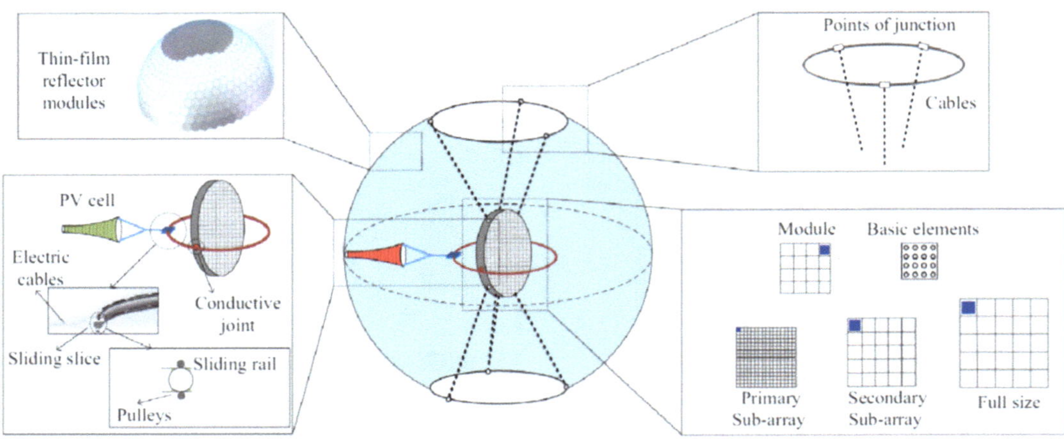

Fig. 3.26 Scheme of SSPS-OMEGA

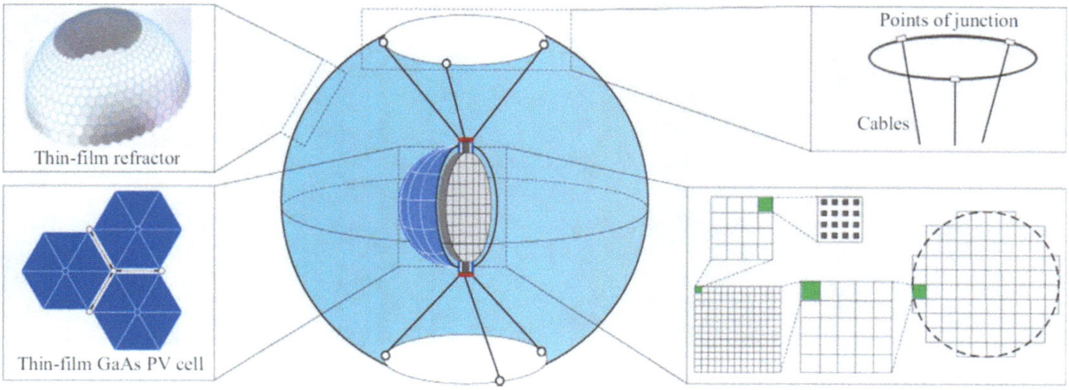

Fig. 3.27 Scheme of improved SSPS-OMEGA

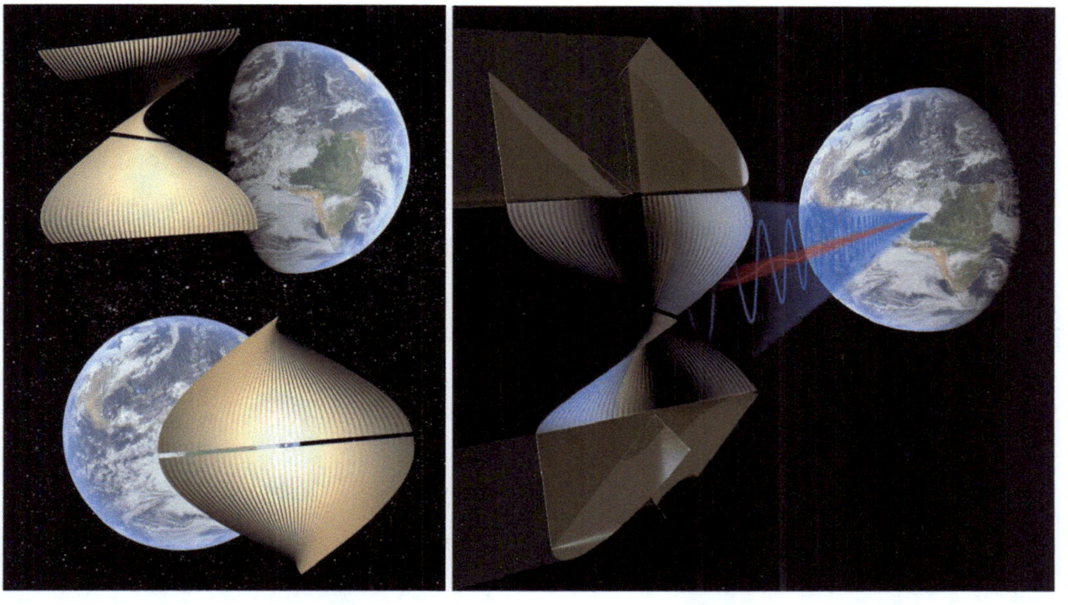

Fig. 3.28 Illustration of CASSIOPeiA SPS (Left: 2017 Version; Right: 2018 Version). (Credit: Ian Cash)

46% under 508 Sun Constant illumination. Through the special transmitting antenna that can control the beam direction in 360°, continuous power transmission to the ground is realized during the orbit period without using rotating parts.

For the 2017 version design, the local concentrating system is installed on the side of each layer, each layer keeps sun-oriented in sequence. The sunlight is incident horizontally and is concentrated on the PV cell module installed on the layered structure surface through the concentrating system, the power generation of each layer changes with different orbit position. For the 2018 version design, a set of reflecting systems is added above and below the entire structure, and the scale of the concentrator is determined according to the concentration ratio. The concentrating system is fixedly connected with the main structure of the SPS. The concentrating system maintains sun orientation, reflects sunlight from the upper and lower directions to the main structure of the SPS, and the reflected light reaches the upper and lower surfaces of each layer structure. The surfaces that can be irradiated by sunlight are all equipped with planar Fresnel concentrators and

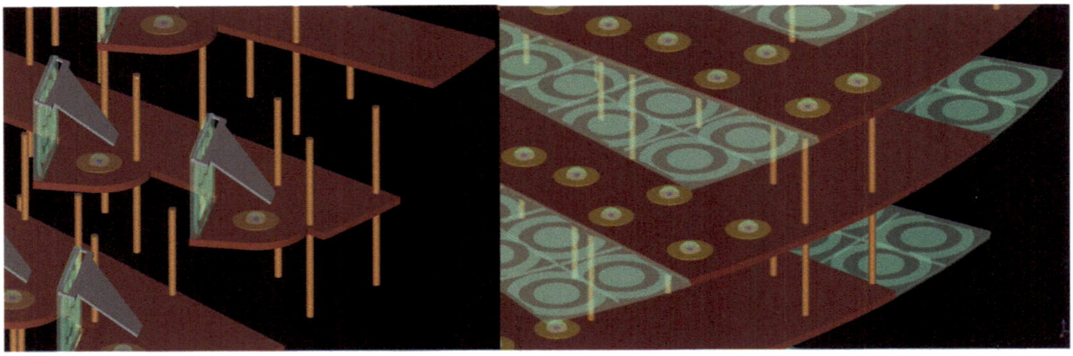

Fig. 3.29 Concentrating PV module and microwave transmitting antenna (Left: 2017 Version; Right: 2018 Version). (Credit: Ian Cash)

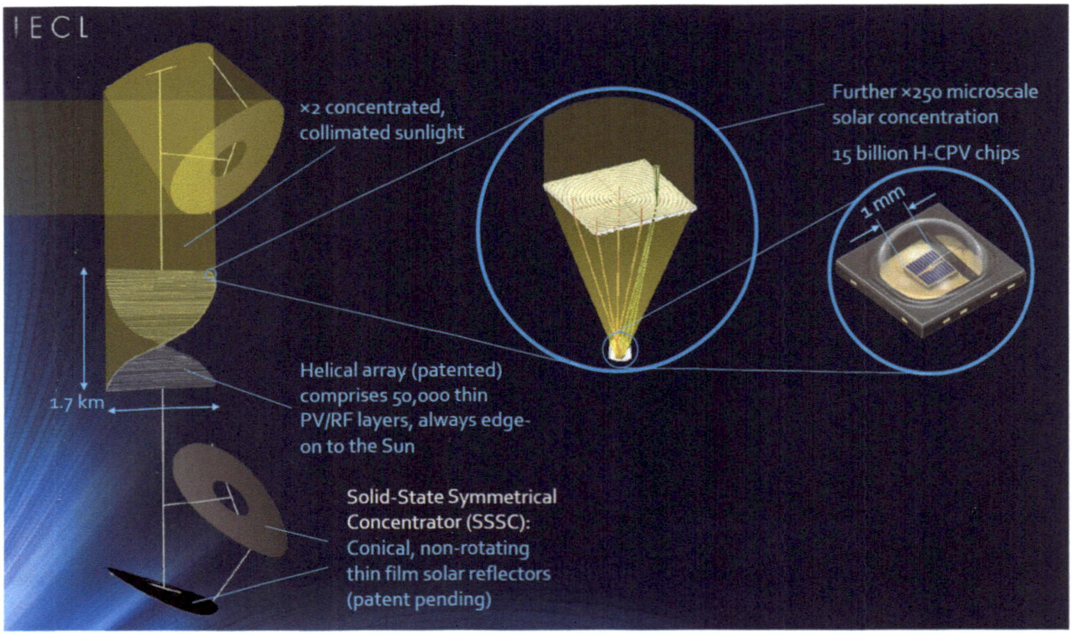

Fig. 3.30 Concept diagram of CASSIOPeiA SPS (2021 Scheme). (Credit: Ian Cash)

concentrating PV cell modules, which can ensure stable power generation, and the generated power can be directly transmitted to the corresponding microwave transmitting antenna (Fig. 3.29). The latest scheme is shown in Fig. 3.30. The corresponding structure's transmitting antenna diameter is 1.7 km, including 50,000 layers of PV/RF combined structure. The fixed concentrating system on the top and bottom can achieve two times concentration. The nominal power generation is 2 GW, and the total mass is 2000 t.

3.5 Laser Power Transmission SPS

3.5.1 LPT SPS

Due to issues such as the atmospheric transmittance, efficiency, and safety of the laser, the proposed high-power SPS concepts are mainly based on microwave power transmission. However, laser power transmission has a narrow beam and

Fig. 3.31 Illustration of LPT SPS. (Credit: SpaceWorks Enterprises/Astrium)

flexible direction control, which is suitable for small-scale power supply applications on the ground and space power transmission applications, especially suitable for special deep space exploration power supply applications such as lunar exploration. Its most important feature is that it can transmit energy separately through multiple dispersed laser transmission systems (Fig. 3.31), directly superimposing energy at the receiving end, without the need to transmit power through a whole microwave transmitting antenna like microwave power transmission, which greatly reduces the complexity of the overall structure and electric power transmission and management.

In 2002, Boeing proposed a concept called Skylight for a laser transmission SPS (Fig. 3.32). This concept consists of 10 parallel laser transmission units. Each laser transmission unit includes a pair of solar arrays used by the International Space Station, a set of lasers and transmission devices, and related platform equipment. The total length of the SPS is 170 m, which can generate 2.7 MW of electricity. Considering a 30% laser conversion efficiency, the total laser power is close to 1 MW. For the concept of a

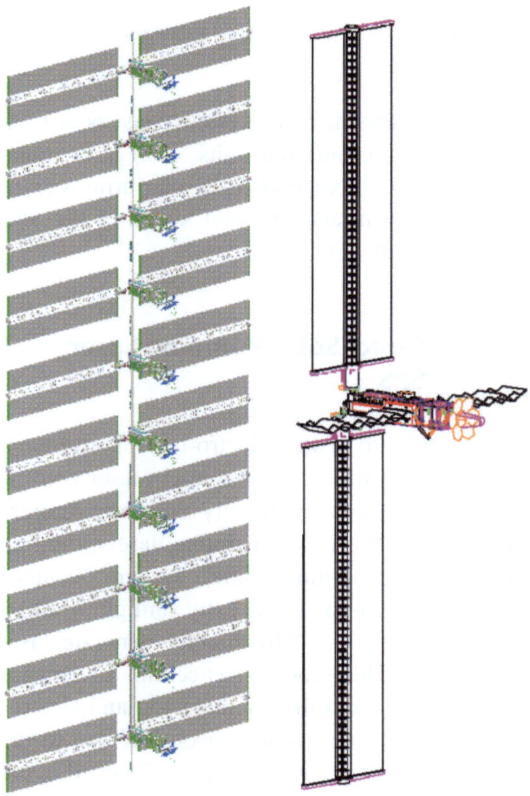

Fig. 3.32 Concept of Skylight LPT SPS

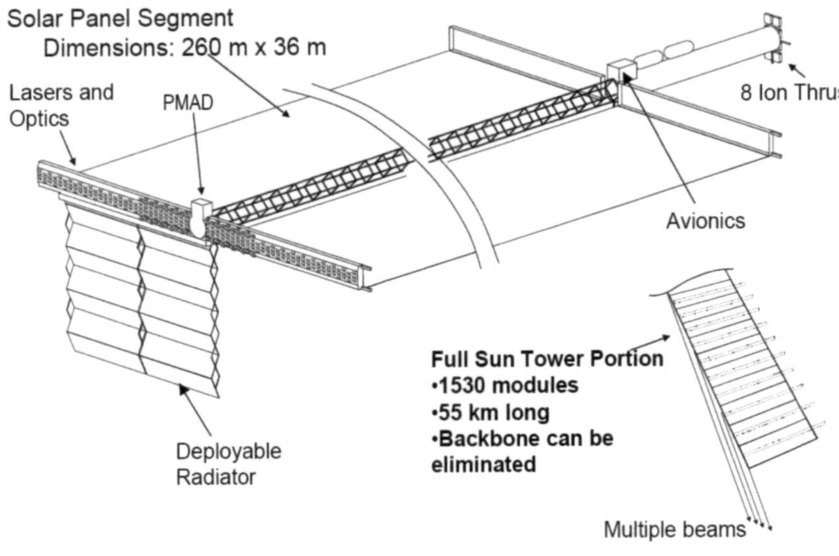

Fig. 3.33 Concept of a GW-level LPT SPS

GW-level SPS, the entire system consists of 1530 large laser transmission units, each laser transmission unit corresponds to a solar array size of 260 m × 36 m, and the total length of the SPS reaches 55 km (Fig. 3.33).

3.5.2 Direct Solar-Pumping Laser SPS

In addition to using electro-laser conversion, lasers also have a special sunlight-laser conversion method, which directly converts sunlight into lasers. Direct solar-pumping lasers, through a high-concentration ratio system, achieve extremely high-power intensity sunlight irradiating the laser crystal to directly generate laser (Fig. 3.34). This scheme has certain advantages because it reduces the photoelectric and electro-optical energy conversion links. But due to the current low conversion efficiency, it is still in the preliminary research stage. The direct solar-pumping laser SPS proposed by JAXA uses a super-large solar concentrator to concentrate sunlight, and the concentrated sunlight irradiates the laser medium, generates lasers using the direct pumping laser method, then the laser is transmitted through the optical system. The receiving end uses a specific laser PV cell that matches the corresponding laser frequency to receive the laser and convert it into electricity efficiently.

Figure 3.35 shows a 10 MW direct solar-pumping laser SPS unit, which consists of a pair of solar concentrators, a laser containing laser medium, a thermal radiator, a laser emission system, and a platform system. The laser medium that can be used by the laser includes sapphire or YAG crystals, and the output laser wavelength is 1064 nm. The solar concentrator is a symmetrical structure. In order to obtain the required solar radiation intensity, its width reaches 200 m, and the height is 100 m. By focusing sunlight through a 100 m long solar-pumping laser, a laser is generated through the laser medium within it, with an expected conversion efficiency of 19%. The power intensity at the focus is very high, and the laser medium needs to use a combination of liq-

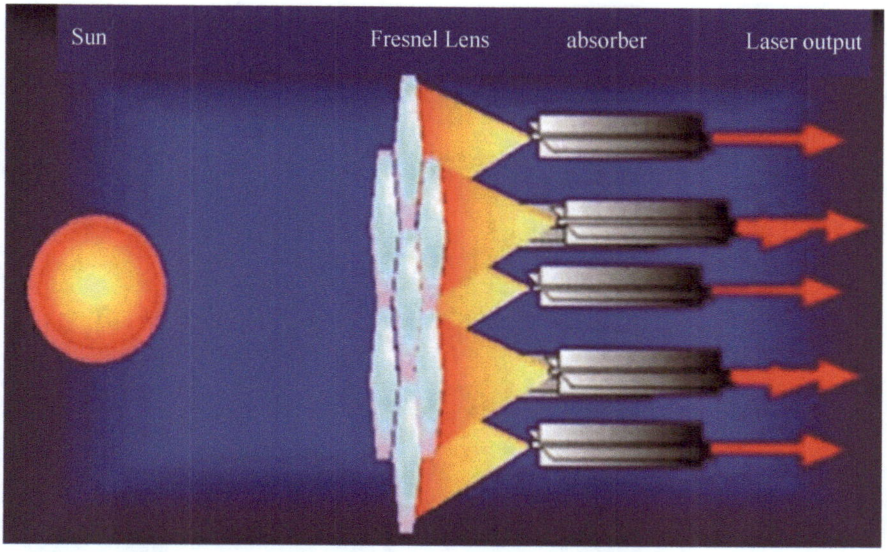

Fig. 3.34 Diagram of direct solar-pumping laser. (Credit: JAXA)

Fig. 3.35 Direct solar-pumping laser SPS unit. (Credit: JAXA)

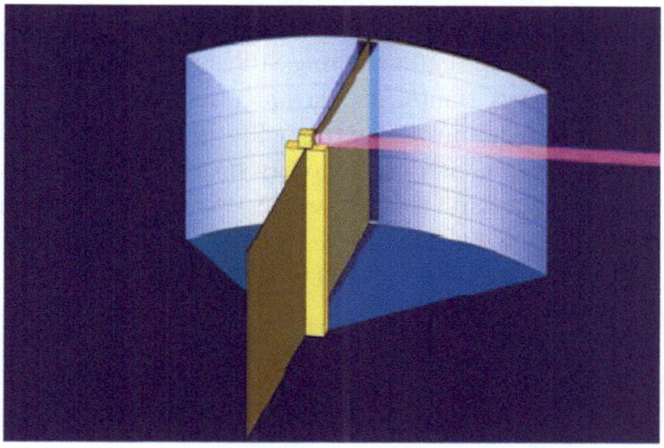

uid cooling and thermal radiation for temperature dissipation. The waste heat generated by the laser is absorbed by the fluid circuit and dissipated through a heat radiator, which measures 100 m × 200 m. Due to the concentration ratio reaching hundreds of times, the conversion efficiency of the laser and the heat dissipation of the system become key factors. The concentrating mirror always maintains sun orientation during orbital operation, and the laser's direction to the receiving end is adjusted by the laser emission system. For a GW-level direct solar-pumping laser Solar Power Satellite (Fig. 3.36), it is composed of 100 parallel 10 MW direct solar-pumping laser SPS units.

Fig. 3.36 Illustration of a GW direct solar-pumping laser SPS. (Credit: JAXA)

3.6 Lunar Solar Power Station

In addition to the earth-orbiting SPS, establishing a solar power station on the lunar surface can generate electricity uninterruptedly 24 h a day, 7 days a week, and continuously provide clean energy to the Earth through wireless power transmission. The concept of a lunar solar power station was proposed by Dr. David R. Criswell of the United States in 1986.

The lunar surface environment is very suitable for large-scale solar power generation. The lunar surface is a vacuum, with stable sunlight conditions; the lunar surface area is huge, allowing for the deployment of a large area of solar panels; the lunar surface is very quiet, so the surface of the solar cells will not be contaminated; one side of the moon always faces the Earth, making it suitable for installing power transmission devices; the moon is rich in resources, suitable for directly using in situ resources for the construction of solar power stations, including in situ production of solar cells (silicon cells), transmission cables, power station structures, etc. The main issues include: considering the relative position of the moon and the sun, solar panels need to be laid out continuously around the lunar equator, using cables to transmit power to the energy emission device (microwave transmitting antenna and laser emitting device) to achieve continuous power transmission; due to the rotation of the Earth, multiple energy receiving stations need to be built on the earth's surface or in earth's orbit to achieve continuous energy reception.

In 2009, Japan's Shimizu Corporation proposed an ambitious plan to build a solar power station on the moon—the "Luna Ring." The plan envisages building a solar power generation belt around the moon's equator (about 11,000 km), then converting the electricity into microwaves or lasers and sending it back to Earth, where the microwave or laser are received on the ground and reconverted into electricity (Fig. 3.37). The initial width of the "Luna Ring" solar power generation belt is on the kilometer scale, which can gradually be expanded to 400 km. The electricity generated by the solar panels is transmitted to multiple energy emission devices facing the Earth via cables, and energy is transmitted to multiple receiving stations on the ground in the form of laser or microwave (the diameter of the microwave transmitting antenna is about 20 km). The "Luna Ring" project is huge, and the plan maximizes the use of in situ lunar resources for construction through various autonomous devices, with some materials needing to be transported from the earth.

Fig. 3.37 Illustration of lunar solar power station. (Credit: Shimizu Co.)

Bibliography

1. The Luna Ring–Lunar Solar Power Generation[R]. Shimizu Corporation, 2009.
2. Choi J M, Moon G W. Conceptual Design of Korean Space solar power Satellite[C]. 70th International Astronautical Congress, Washington D. C., USA, 2019.
3. Cash I. CASSIOPeiA—A New Paradigm for Space solar power[C]. 69th International Astronautical Congress, Bremen, Germany, 2018.
4. Mankins J C. SPS-ALPHA Mark-III and an Achievable Roadmap to Space solar power[C]. 71th International Astronautical Congress, Dubai, UAE, 2021.
5. Mankins J C. SPS-ALPHA: Evolving Markets, Capabilities and Concepts of Operations for Modular & Practical Space solar power (SSP) [C]. 72th International Astronautical Congress, Paris, France, 2022.
6. Sasaki S, Tanaka K, Higuchi K, et al. Feasibility Study of Multi-Bus Tethered-SPS[C]. 59th International Astronautical Congress, Glasgow, UK, 2008.
7. Arya M, Lee N, Pellegrino S. Ultralight Structures for Space solar power Satellites[C]. 3rd AIAA Spacecraft Structures Conference, San Diego, USA, 2016.
8. Seboldt W, Klimke M, Leipold M, et al. European Sail Tower SPS Concept[J]. Acta Astronautica, 2001, 48: 785-792.
9. Mankins J C. A Fresh Look at Space solar power: New Architectures, Concepts and Technologies[J]. Acta Astronautica, 1997, 41: 347-359.
10. Mankins J C. A Technical Overview of the SUNTOWER Solar Power Satellite Concept[J]. Acta Astronautica, 2002, 50: 369-377.
11. Sasaki S, Tanaka K, Higuchi K, et al. A New Concept of Solar Power Satellite: Tethered-SPS[J]. Acta Astronautica, 2006, 60: 153-165.
12. Hou X B. Study on Multi-Rotary Joints Space Power Satellite Concept[J]. Aerospace China, 2018, 19 (1): 19-26.
13. Cash I. CASSIOPeiA – A New Paradigm for Space solar power[J]. Acta Astronautica, 2019, 159: 170-178.
14. YangY, Zhang Y Q, Duan B Y, et al. A Novel Design Project for Space solar power Station (SSPS-OMEGA) [J]. Acta Astronautica, 2016, 121: 51-58.
15. Hou X B, Wang L, Zhang X H, et al. Concept Design on Multi-Rotary Joints SPS[J] (in Chinese). Journal of Astronautics, 2015,36(11):1332-1338.
16. Yang Y, Duan B Y, Huang J, et al. SSPS-OMEGA: A New Concentrator System for SSPS[J] (in Chinese). Chinese Space Science and Technology, 2014, (5):18-23.
17. Hou X B, Zhang X, Cheng Z A, et al. Concept Design on Modular Multi-Rotary Joints Solar Power Satellite[J] (in Chinese). Space Electronic Technology, 2023, 20(3):1-6.

High-Power Space Solar Power Generation System

4

4.1 Space Solar Power Generation Methods

Space solar power generation technology is one of the main factors affecting the overall efficiency, volume, and weight of the entire SPS and is the decisive factor for the cross-sectional area of the SPS. There are several ways to generate power in space, as shown in Fig. 4.1.

1. Solar photovoltaic power generation
   Solar PV cells directly receive sunlight and generate electricity through photovoltaic effects, which is the most widely used power generation method in space applications. Currently, the power generation efficiency of solar PV cells used in space has exceeded 30%.
2. Solar concentration photovoltaic power generation
   By using a concentrator to gather sunlight and irradiate solar cells to generate electricity, the area of solar cells can be reduced and the power generation efficiency of the cells can be improved. This method can be used for special power system design and has been applied in space.
3. Solar concentration thermodynamic power generation
   By using a concentrator to gather sunlight onto a heat absorber, and then driving a heat engine to generate electricity through the working medium of the heat absorber. Brayton or Stirling heat engines are mainly considered using for power generation in space, which have high-power generation efficiency. Combined with high-temperature heat storage, power supply during the shadow period can be achieved, as shown in Fig. 4.2. This method was studied as the main power supply scheme for the US Freedom Space Station in the 1990s, and a 2 kW ground prototype was developed. However, due to the complexity of the system and the presence of long-term moving parts, its reliability needs to be verified. Later, with the advancement of photovoltaic power generation technology and energy storage technology, it did not get practical application.

X. Hou, *Space Solar Power*, Sustainable Development Goals Series,
https://doi.org/10.1007/978-981-97-3580-8_4

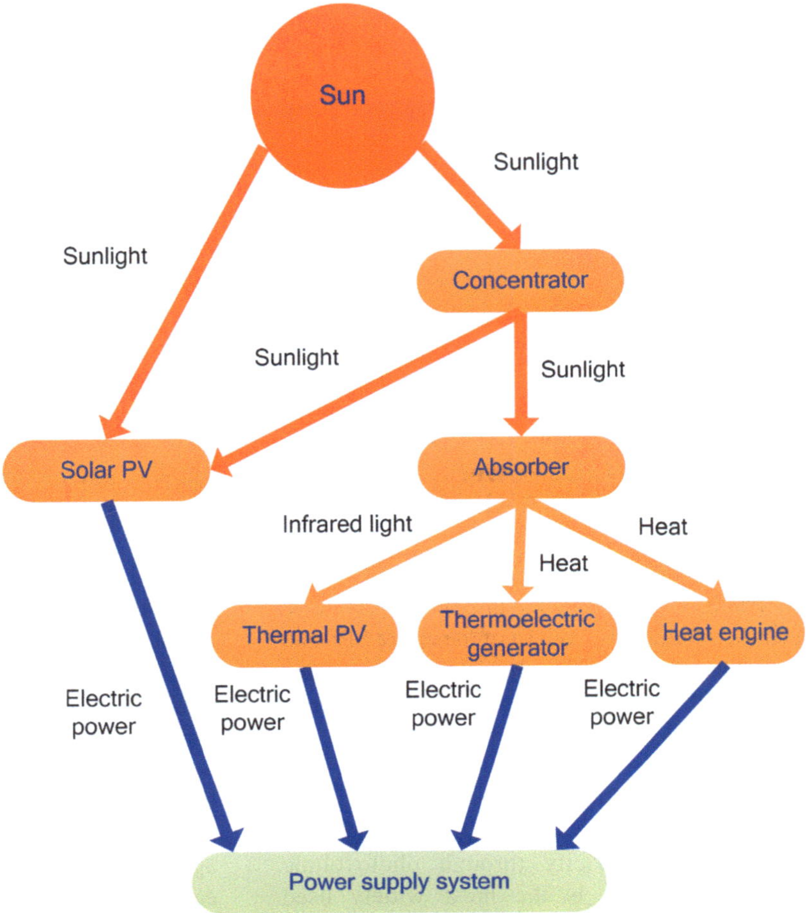

Fig. 4.1 Different space solar power generation methods

4. Solar concentration thermoelectric power generation
   By using a concentrator to gather sunlight onto a heat absorber, and then relying on the temperature difference at both ends of the thermocouple thermoelectric battery, electricity is generated using the Seebeck effect. Currently, the thermoelectric power generation method is widely used in isotope nuclear power sources of deep space probes, such as the Perseverance Mars Rover, which uses PbTe (lead telluride)/TAGS (tellurium, silver, germanium, antimony composite material) medium temperature zone thermoelectric materials, with hot end temperature and cold end temperature of 550 and 165 °C, respectively, as shown in Fig. 4.3.
5. Solar concentration thermophotovoltaic power generation
   Use a concentrator to gather sunlight onto a heat absorber to generate high temperatures above 1000 °C. Then thermophotovoltaic cells is used to absorb high-temperature infrared radiation (0.8–2 μm) for photovoltaic power generation. The cell materials selected are GaSb and InGaAs, and the power generation efficiency is expected to reach 20%, as shown in Fig. 4.4. Currently, this power generation method has not been applied in space.

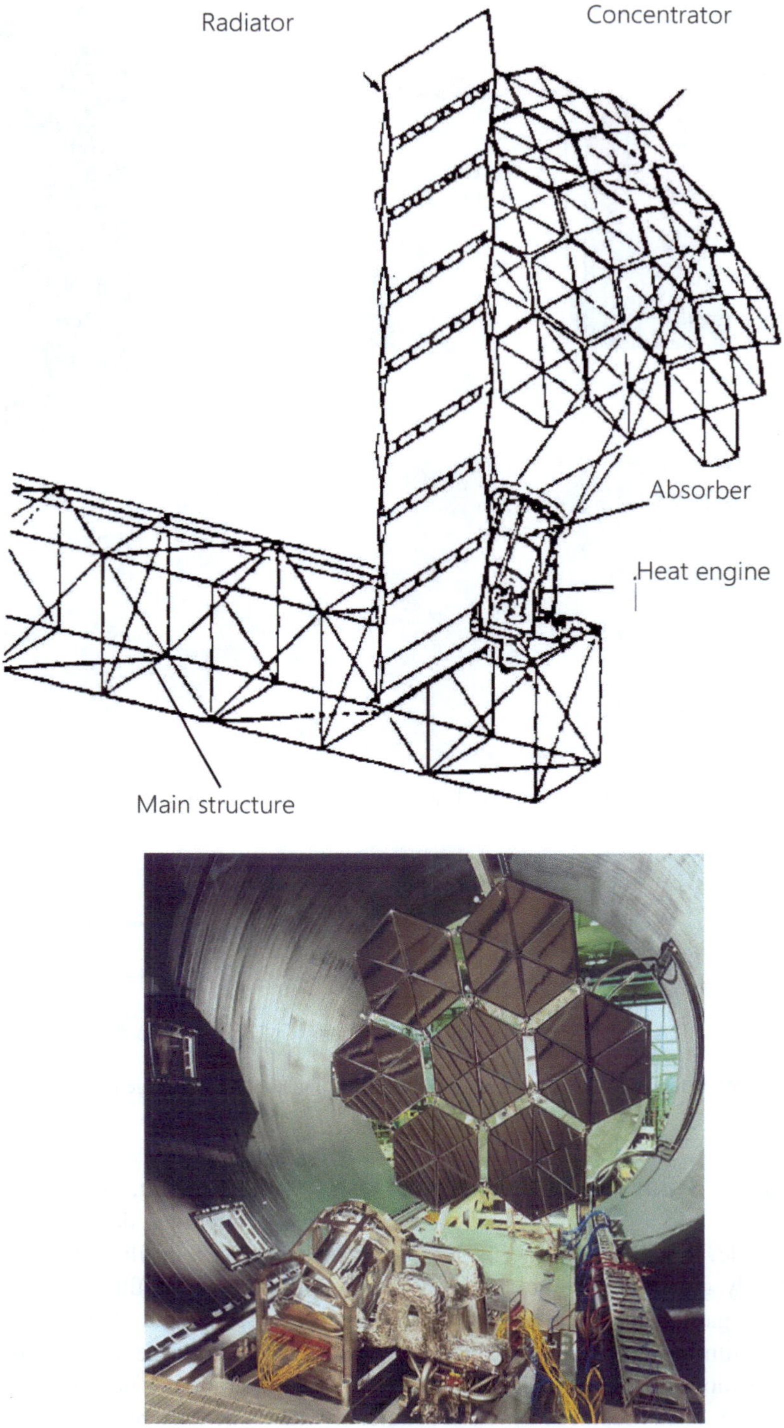

Fig. 4.2 Solar concentrating thermodynamic power generation. (Credit: NASA)

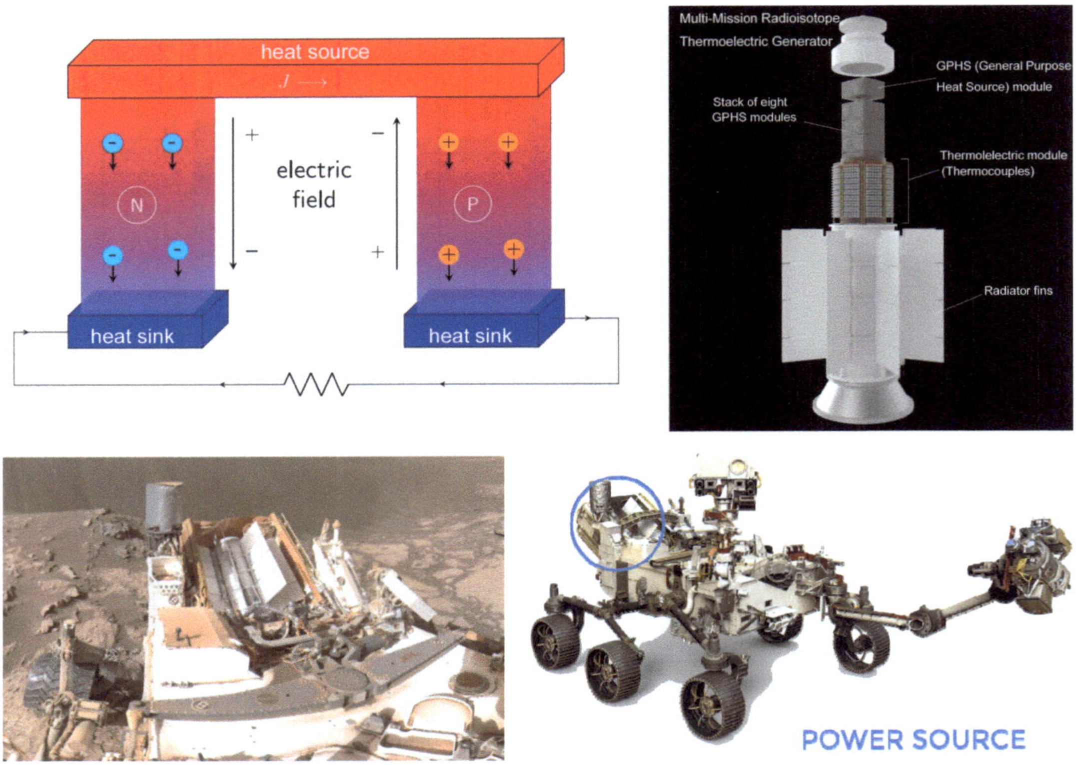

Fig. 4.3 Seebeck effect, isotope nuclear power source structure, the perseverance Mars rover. (Credit: NASA)

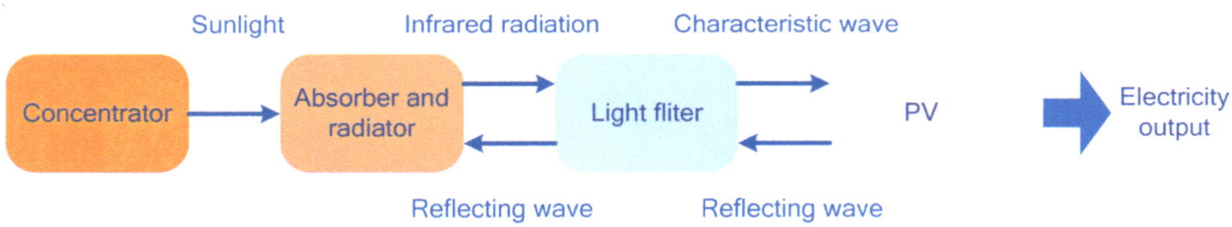

Fig. 4.4 Schematic diagram of solar concentration photovoltaic power generation

4.2 Solar PV Cells for Space Use

The development of space solar PV cells has mainly gone through the stages of silicon solar cells, gallium arsenide (GaAs) solar cells, and thin-film solar cells. Early spacecraft generally used monocrystalline silicon solar cells, which were gradually replaced by GaAs solar cells with high efficiency and good space radiation tolerance. The most widely used currently is the triple-junction GaAs solar cell and the conversion efficiency on-orbit has exceeded 30%. With the demand for high-power generation by large spacecraft, thin-film solar cells have become an important development direction, mainly including thin-film silicon cells, CIGS thin-film cells, and thin-film GaAs cells. Concentrating solar cells have also been developed because they can reduce the quanlity of cells, improve power generation efficiency, and are suitable for implementing high-voltage arrays. To achieve higher efficiency, smaller mass, and lower cost, the main development directions of space solar PV cells include multiple-junction GaAs solar cell, thin-film GaAs solar cell, and the space application of perovskite solar cells.

1. Rigid solar cells
   (a) Silicon solar cell. Silicon solar cell was the earliest developed solar cell in the world. It is widely used in terrestrial solar power generation and early space applications, currently mainly used for low-cost commercial satellites in space. Silicon solar cells can be divided into three types: monocrystalline silicon solar cell, polycrystalline silicon solar cell, and amorphous silicon solar cell. At present, the highest photoelectric conversion efficiencies of monocrystalline silicon and polycrystalline silicon solar cells are 26.1% and 24.2%, respectively.
   (b) GaAs solar cell. Different semiconductor materials have different band gaps, as shown in Fig. 4.5. GaAs solar cell is a type of compound cell, belonging to the III—V compound cells. GaAs has a band gap width of 1.42 eV, which matches well with the solar spectrum and has advantages such as high photoelectric conversion efficiency, high-temperature resistance, and space radiation tolerance. Since single-junction GaAs solar cell only responds to a specific narrow band of the spectrum, their photoelectric conversion efficiency is limited, with the highest conversion efficiency currently achieved being 27.8%.
   (c) Multi-junction GaAs solar cells. In order to achieve higher efficiency, it is necessary to utilize a wider range of the solar spectrum. This requires targeting different spectral segments, increasing the number of sub-cells for precise spectrum matching, and stacking the P-N junctions of different semiconductor materials in an ascending order from bottom to top according to the band gap width. This allows high-energy photons to be absorbed in the upper wide band gap junction and low-energy photons to be absorbed in the lower narrow band gap junction, effectively improving the conversion efficiency of the solar cell. Figure 4.6 shows the typical structure of a III-V triple-junction gallium arsenide cell (GaInP/InGaAs/Ge), where GaInP, InGaAs, and Ge are matched to the 300–650 nm, 650–950 nm, and 950–1800 nm spectrum segments, respectively, to make the best use of the solar spectrum for efficient photoelectric conversion, with the highest laboratory photoelectric conversion efficiency currently achieved being 39.5%. The future will move towards the development of four-junction and above technologies, but the more junctions, the more complex the manufacturing process, and the higher the requirements for structural matching, which poses great technical difficulties for practical engineering applications.

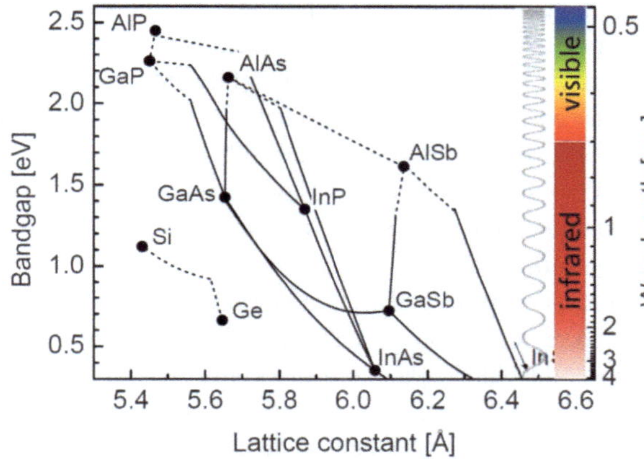

Fig. 4.5 Band gaps of different semiconductor materials

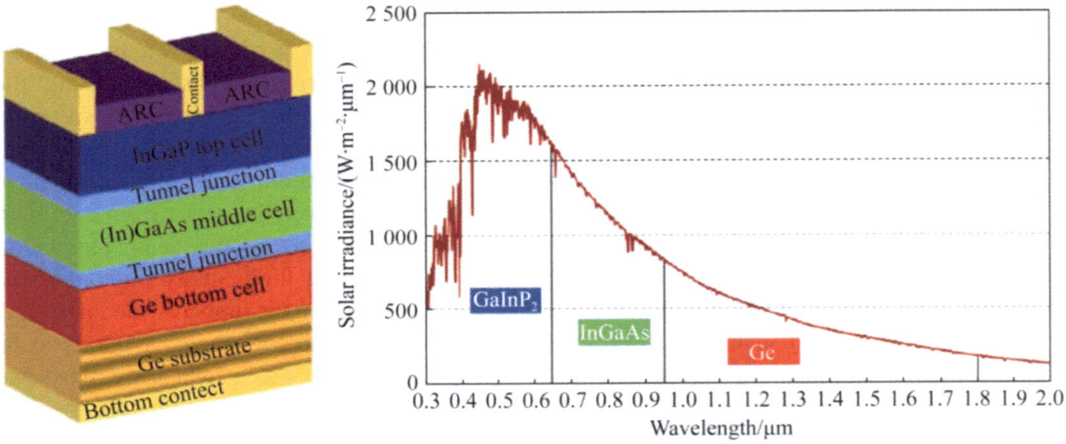

Fig. 4.6 Structure and absorption spectrum of triple-junction GaAs cell

Fig. 4.7 IKAROS solar sail and its thin-film solar cells. (Credit: ISAS)

2. Thin-film solar cells

Traditional space solar cells are rigid cells. To meet the requirements for folding and unfolding of large solar arrays and lightweight, it is necessary to develop thin-film solar cells with good bending performance. The types of flexible solar cells currently developed mainly include crystalline silicon thin-film solar cell, amorphous silicon thin-film solar cell, copper indium gallium selenium (CIGS) thin-film solar cell, CdTe thin-film solar cell, GaAs thin-film solar cell, perovskite solar cells, etc. Considering factors such as technological maturity, efficiency, and cost, thin-film solar cells have not yet been widely used in space. Among them, amorphous silicon thin-film solar cells were used in Japan's IKAROS solar sail demonstration mission (Fig. 4.7). In the near to medium future, the most likely flexible solar cells to be widely used in space include CIGS thin-film solar cells and GaAs thin-film solar cells, and perovskite solar cells may also be practically used in space as the technology matures.

(a) CIGS thin-film solar cells. Copper indium gallium selenium is a I—III—VI group quaternary compound semiconductor, which is a mixed crystal semiconductor of $CuInSe_2$ and $CuGaSe_2$. By changing the content of Ga and In elements, their bandgap values can be continuously adjusted within the range of

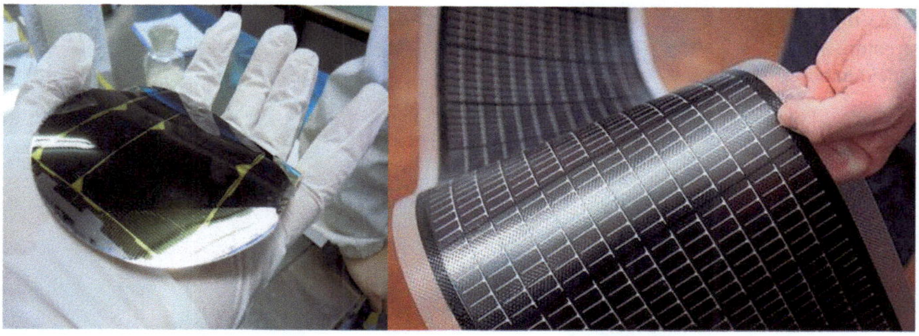

Fig. 4.8 CIGS thin-film solar cells

1.02–1.67 eV. The typical structure of a CIGS thin-film solar cell is a substrate (Mo), a back electrode (CIGS), an absorption layer (CdS), a buffer layer (i-ZnO layer/TCO layer/Ni-Al-Ni), and a metal grid electrode. CIGS thin-film solar cells have the advantages of high efficiency, high absorption coefficient, strong radiation tolerance, good weak light performance, and low cost. The theoretical conversion efficiency is greater than 30%, and the highest conversion efficiency achieved so far is 26.7%. Because CIGS cells can be prepared using a continuous roll-to-roll process based on printing technology, it can achieve industrial production and significantly reduce manufacturing costs, while also facilitating the production of large-area cells, reducing the difficulty of manufacturing cell modules (Fig. 4.8).

(b) GaAs thin-film solar cells. GaAs cells generally use germanium as the substrate, with a substrate thickness of 145–175 μm, while the thickness of the triple-junction GaAs cell itself is only about 10 μm. The germanium substrate determines the rigidity of the entire GaAs cell. Therefore, the core of the GaAs thin-film solar cell is to replace the thick rigid substrate with a thin flexible substrate, currently mainly using polyimide film substrates, and a special inverted manufacturing process is required. The main preparation process is: first grow a GaInP top cell on the GaAs (or Ge) substrate, followed by the middle cell and the bottom cell, then transfer the epitaxial layer to the polyimide film via bonding technology, then use a chemical method to remove the original GaAs substrate, expose the top of the solar cell, and finally optimize the flexible cell device process to obtain a high-efficiency lightweight flexible GaAs thin-film solar cell (Fig. 4.9). At present, the conversion efficiency of the GaAs thin-film solar cells produced has exceeded 30% (AM0, 25 °C), and the power-to-weight ratio has reached over 2000 W/kg.

3. Concentrating solar cells

The initial intention of researching concentrating solar cells was to reduce the cost of power generation system by using low-cost concentration systems to reduce the amount of solar cells used. Since the solar radiation intensity on the surface of the concentrating solar cell reaches several times the solar constant, the direct effect is that the short-circuit current of the solar cell increases linearly with the concentration ratio, while the open-circuit voltage increases with the concentration ratio in a near-logarithmic relationship, and the cell conversion efficiency first increases and then decreases with the increase of the concentration ratio, mainly due to the influence of internal resistance. Therefore, the design of concentrating cells should minimize resistance loss and achieve higher conversion efficiency by structural optimization design. Currently, the highest photoelectric conver-

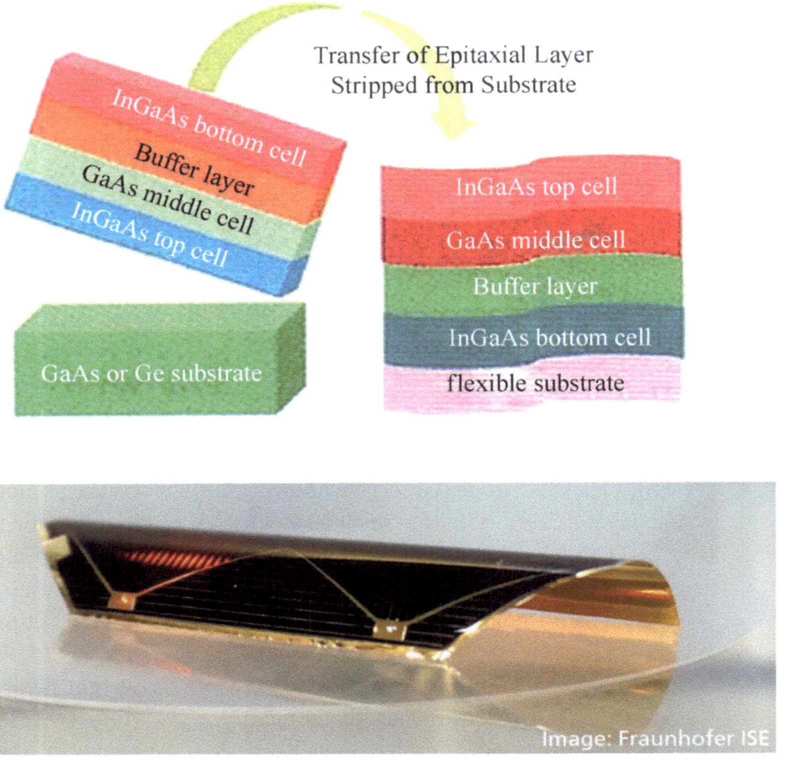

Fig. 4.9 GaAs thin-film solar cell structure and cell. (Credit: Fraunhofer ISE)

sion efficiency of triple-junction GaAs concentrating cells has reached 44.4% (4.9% higher than non-concentrating cells), and the highest laboratory photoelectric conversion efficiency of six-junction GaAs concentrating cells has reached 47.1%. The high efficiency of multi-junction GaAs concentrating cells makes it have a good application prospect in space. Since concentrating solar cells need to work under very high solar radiation intensity, effective heat dissipation of the cell is one of its key technologies.

4.3 Space Solar Array

4.3.1 Typical Space Solar Array

The typical space solar array mainly includes body-mounted, rigid deployable, semi-rigid deployable, and flexible deployable types. The current direction of technological development is towards high-power, long-life, high-power-to-weight ratio, high-power-to-area ratio, low stowage volume, and concentrating solar arrays are also one of the important development directions.

The body-mounted solar array is a configuration where solar cells are directly deployed on the surface of the satellite. This structure is simple, but power expansion is limited, making it only suitable for spacecraft with lower output power requirements (Fig. 4.10a).

The rigid deployable solar array is the main type of solar array currently used in spacecraft. It is constructed by deploying rigid solar cells and circuits on a rigid substrate (usually an aluminum honeycomb), typically composed of multiple subarray structures. During the launch phase, it is folded and compressed on the surface of the satellite body, and after entering orbit, it is deployed in one or two dimensions to form a large-area solar array (Fig. 4.10b). The structure of the rigid solar array is simple and has a high rigidity, which is convenient for achieving high-power generation through subarray expansion. However,

(a) Body-Mounted Solar Array

(b) Rigid Deployable Solar Array

(c) Semi-Rigid Substrate Solar Array

Fig. 4.10 Typical applications of different solar arrays. (**a**) Body-mounted solar array, (**b**) rigid deployable solar array, (**c**) semi-rigid substrate solar array. (Credit: NASA)

its overall mass and stowage volume are large, resulting in a lower power-to-weight ratio.

The semi-rigid substrate is a modification of the rigid solar array, where the rigid substrate is replaced with a substrate composed of a carbon fiber reinforced composite material frame and a fiber woven mesh, with rigid cells deployed on the fiber woven mesh (Fig. 4.10c). Without affecting the deployment and mechanical properties, it can reduce the overall mass of the solar array and improve the power-to-weight ratio. The semi-rigid substrate has been successfully applied in the "Tiangong" space laboratory and DFH-5 communication satellite launched by China.

For the application needs of ultra-high-power spacecraft such as space stations, in order to meet the requirements of high-power, high-power-to-weight ratio, and high stowage-to-deployment ratio for ultra-large solar arrays, the development of flexible solar arrays is inevitable. Flexible solar arrays use flexible substrates, adhere rigid or thin-film solar cells and flexible circuits, and have the advantages of small stowage envelope, light weight, and high-power-to-weight ratio. They are the future development trend of ultra-high-power spacecraft.

4.3.2 Space Concentrating Solar Array

The concentrating solar array is based on concentrating solar cell technology, which uses a concentrator to form a high concentration ratio of incident sunlight on the solar cell, solar array module, or solar array through point concentration, line concentration, or surface concentration. This can enhance the efficiency and utilization of space solar cells, reduce system costs, and achieve special spacecraft configurations and high voltage power output. Considering different concentrating method of the concentrator, it can mainly be divided into reflective and refractive types. The concentrator structures currently being researched mainly include V-shaped reflective structure, parabolic reflective structure, Fresnel lens refractive structure, and integral reflective structure.

For the V-shaped reflective concentrating solar array, the concentrator is initially compressed on the plane, and after unfolding, it forms the required shape, focusing sunlight onto the solar array module in the middle. Considering the heat dissipation factor, the concentration ratio is two to four times. The newly developed V-shaped reflective concentrator solar array (Fig. 4.11) is being validated on the solar array of the DART satellite in the United States. The main difficulty of the V-shaped reflective concentrator solar array technology lies in the adaptability of the concentrator to the space environment and the maintenance of its shape, which is very sensitive to the direction of sunlight incidence (Fig. 4.12).

The representative of the parabolic reflective concentrating solar array is the new flexible solar array developed by Caltech for the Solar Power Satellite, as shown in Fig. 4.13. This module consists of multiple parabolic concentrators, each of which uses a line concentrating method to concentrate sunlight onto the concentrating solar cell strip on the back of the previous concentrator to achieve a high-efficiency lightweight solar array. Because this scheme requires very high surface accuracy of the concentrator, it is difficult to implement, and the Caltech team has planned to use a planar thin-film solar array.

The Fresnel lens refractive concentrating solar array uses an optimized design of flat, cylindrical, or spherical high-transmittance Fresnel lenses to focus sunlight onto solar cell strips or individual solar cells to achieve high-efficiency

Fig. 4.11 V-Shaped reflective concentrating solar array of dart. (Credit: NASA)

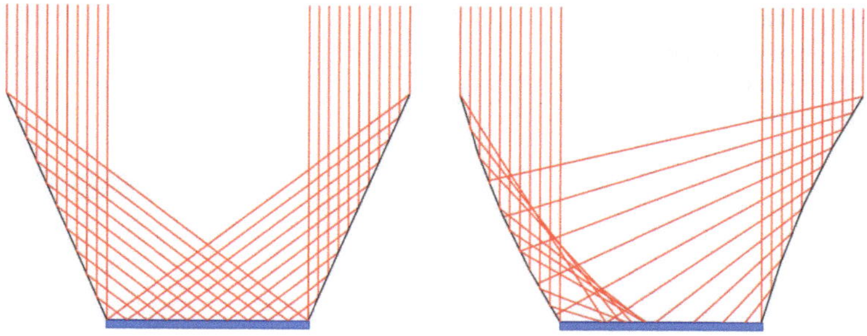

Fig. 4.12 Impact of concentrator deformation on concentrating performance

Fig. 4.13 Parabolic reflective concentrating solar array developed by Caltech. (Credit: Caltech)

power generation (Fig. 4.14). It has been successfully applied on the "Deep Space 1" (DS-1) spacecraft in the United States. The Stretched Lens Array (SLA) developed by the American ENTECH company is a representative scheme, and the cell unit and solar array prototype are shown in Fig. 4.15. The Fresnel lens is initially in a compressed flat state, and after release, it forms the required concentrating structure to achieve line concentration, with a lens thickness of 14 μm. The advantages of this scheme include: it can achieve a high concentration ratio of more than 10 times, if designed for point concentration, it can achieve a higher concentration ratio; it can tolerate large deviations in incident solar radiation; the entire bottom surface can be used as a heat dissipation surface for the solar cell strip, with good heat dissipation conditions; the large spacing between cells can reduce the risk of high voltage discharge and can achieve kilovolt (kV) level high voltage output. The long-term working performance of the Fresnel lens in the space environment is an important factor affecting the application of this technology.

The overall reflective concentrating solar array refers to the use of multiple reflectors to form a large parabolic, spherical, or other shape of the concentrator system, focusing sunlight onto the solar array surface, forming a complete concentrating solar array. Concentration is an important type in the design of Solar Power Satellite and typical representatives include the Two-Stage Reflection Modular Sandwich Microwave SPS (Fig. 4.16), SPS-ALPHA, SSPS-OMEGA and CASSIOPeiA. Among them, the first two concepts respectively use parabolic and special concentrator systems in conjunction with planar solar arrays, SSPS-OMEGA uses a spherical concentrator system in conjunction with a spindle-shaped solar array (Fig. 4.17), while CASSIOPeiA comprehensively uses planar reflective concentrators, planar Fresnel lenses, and spherical Fresnel lenses to achieve high concentrating cell units. The technical difficulties of the concentrating solar array include high uniform illumination concentrator system and heat dissipation of the high concentration ratio solar array. Figure 4.18 shows the concentrating characteristics of the Two-Stage Reflection Modular Sandwich Microwave SPS at different orbit positions, showing a large illumination non-uniformity. To minimize the non-uniformity of

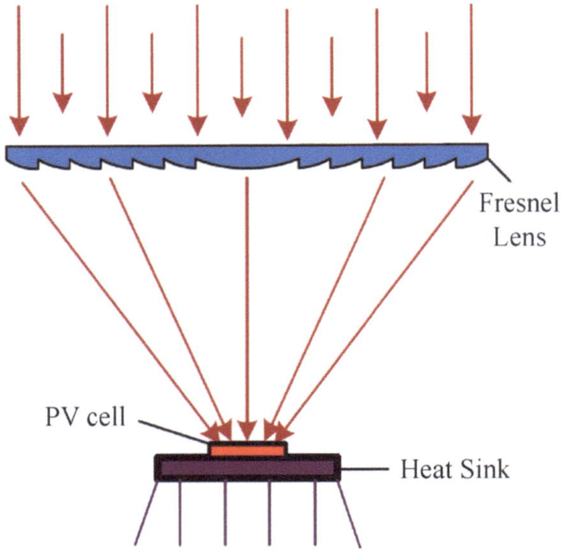

Fig. 4.14 Schematic diagram of Fresnel lens refractive concentrating solar cell

Fig. 4.15 Stretched lens array cell unit and solar array. (Credit: Entech Co.)

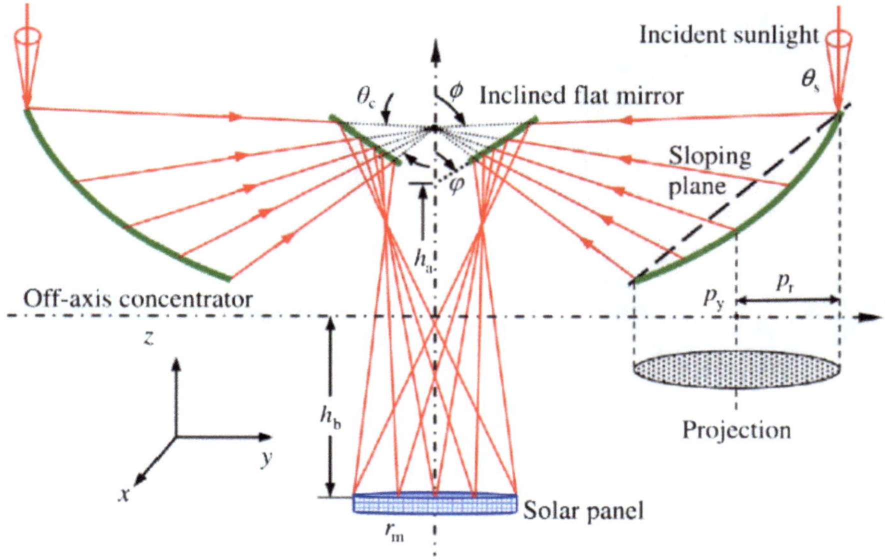

Fig. 4.16 Schematic diagram of two-stage reflection modular sandwich microwave SPS

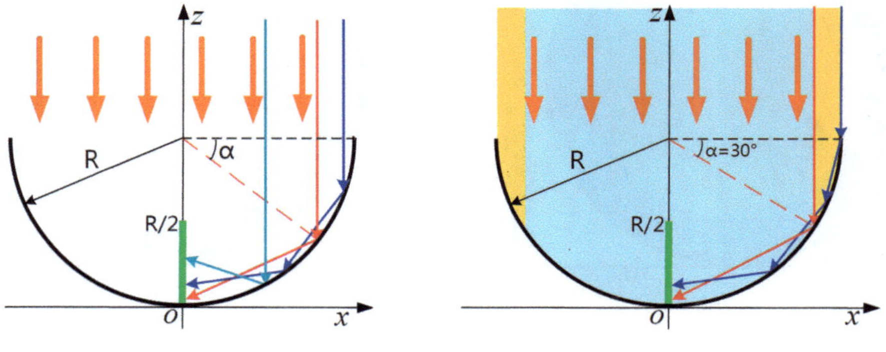

Fig. 4.17 Schematic diagram of spherical reflective concentration system

solar cell surface illumination and improve heat dissipation performance, the spindle-shaped solar array configuration was adopted (Fig. 4.19) for SSPS-OMEGA, and a spectral selective coating was designed to reduce the absorption of useless sunlight by the solar array (Fig. 4.20).

4.3.3 Folding and Unfolding Form of the Flexible Solar Array

For large flexible solar arrays, mechanisms such as mechanical driver unfolding, inflation unfolding, and elasticity unfolding are generally used, and different folding and unfolding forms need to be designed according to the structure of the solar array. The folding and unfolding forms currently being studied mainly include one-dimensional folding and unfolding, fan-shaped folding and unfolding, roll-out deployment, and two-dimensional folding and unfolding.

1. One-dimensional folding and unfolding
   One-dimensional folding and unfolding can also be referred to as Z-shaped folding and unfolding. The corresponding flexible solar array is generally slender, composed of multiple folding subarrays, and is folded in a

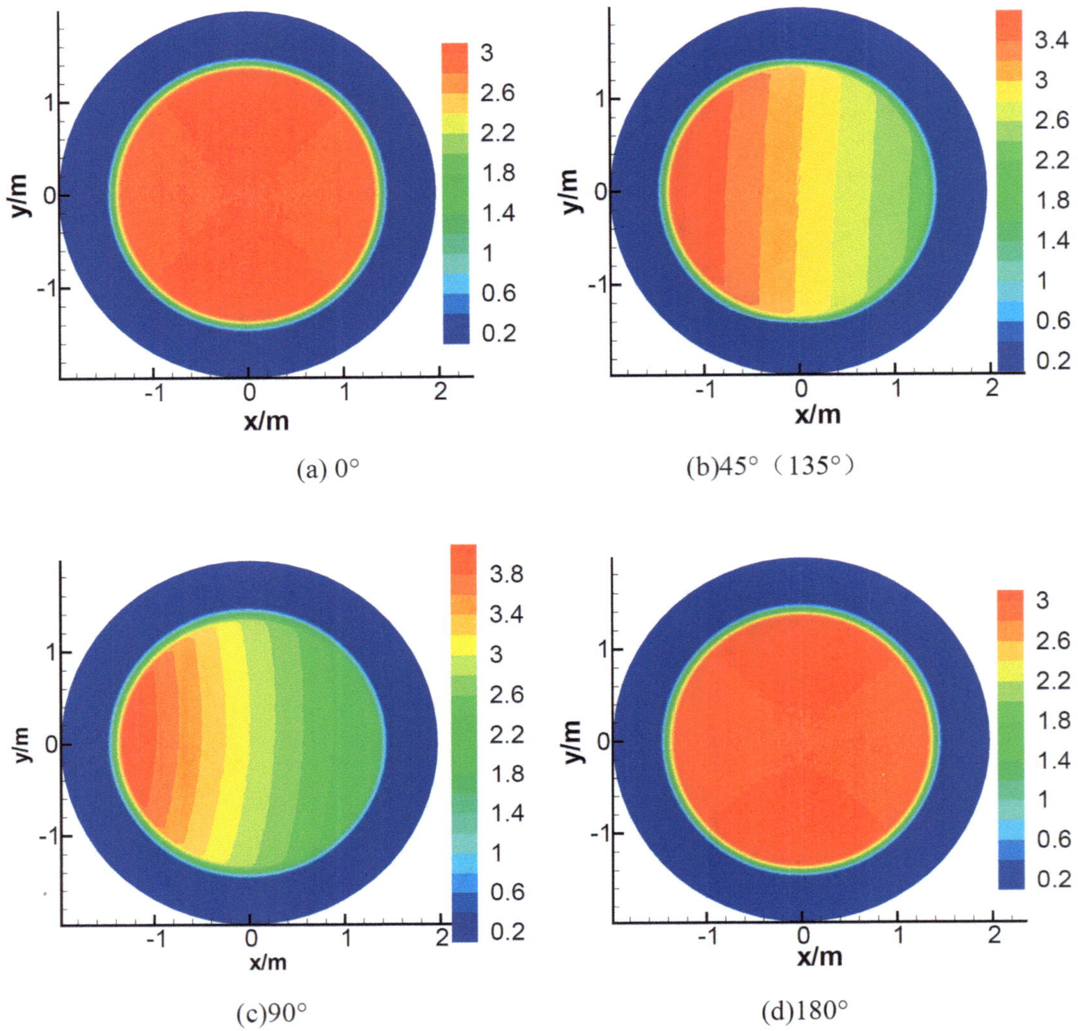

Fig. 4.18 Concentrating characteristics of two-stage reflection modular sandwich microwave SPS at different orbital positions. (**a**) 0°, (**b**) 45° (135°), (**c**) 90°, (**d**) 180°

Z-shape along the long edge direction. After entering orbit, it is unfolded in one dimension along the long edge direction and locked by a mechanism to form a large flexible solar array. Due to the one-dimensional unfolding method, the width direction is strictly limited by the height of the payload fairing, while the length direction can achieve a longer solar array structure according to the folding size constraints. The most typical application is the solar array used by the International Space Station, as shown in Fig. 4.21, which uses silicon solar cells with polyimide film substrate, and uses the motor-driven truss in the middle to unfold the solar array, with a single array unfolding length exceeding 70 m.

2. Fan-shaped folding and unfolding

Fan-shaped folding and unfolding can be regarded as a one-dimensional folding and unfolding form in the circumferential direction. The fan-shaped flexible solar array is circular, decomposed into multiple foldable fan blade subarrays, and folded in a Z-shape along the circumferential direction. After entering

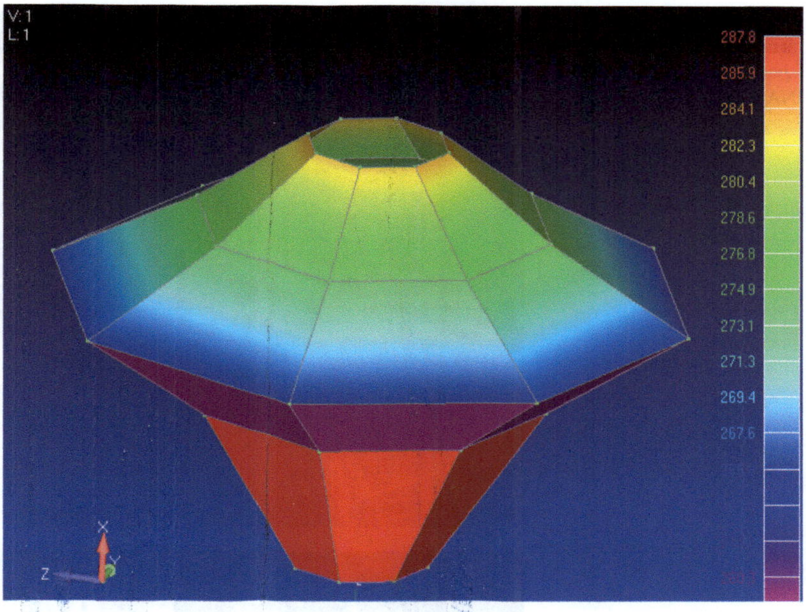

Fig. 4.19 Surface temperature distribution of spindle-shaped solar array

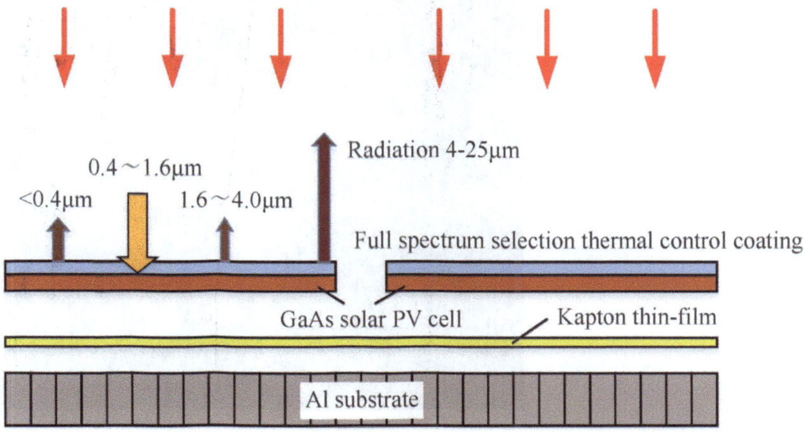

Fig. 4.20 Schematic diagram of spectral selection coating

orbit, the fan blades are unfolded along the circumferential direction through a driving mechanism, and locked by a mechanism to form a circular flexible solar array. The most typical is the flexible solar array UltraFlex developed by ATK Space Systems in the United States, as shown in Fig. 4.22, which consists of 10 groups of foldable triangular thin-film solar subarrays, supported by multiple radial composite material rods. The subarray is a flexible substrate covered with rigid cells. Each group of subarrays is folded in half at the center in the folded state, and 10 groups of subarrays are compressed on the solar array structure. After entering orbit, it is unfolded and locked to form a flexible planar solar array. This scheme has been successfully applied to multiple Mars rovers such as Phoenix and missions like the "Cygnus" spacecraft. The new generation MegaFlex large flexible solar array (Table 4.1) is currently being studied, with a prototype diameter of 9.7 m, a power generation capacity of about 17 kW, an extended diameter of up to 30 m, and it will support future spacecraft missions of tens of kilowatts to hundreds of

Fig. 4.21 Flexible solar array of international space station. (Credit: NASA)

kilowatts. To reduce the folding length, this solar array will be folded again in the folding state.

3. Roll-out deployment

Roll-out deployment involves rolling up the thin-film solar array along the long edge, and then deploying it in a one-dimensional roll-out manner through a drive mechanism after reaching the orbit. The flexible solar array of the Hubble Telescope uses a deployment method from the central axis to both sides (Fig. 4.23). In response to the demand for high-power solar arrays, the American Deployable Space System company designed, developed, and tested a roll-out solar array unit called ROSA (Roll Out Solar Array). The prototype of the solar array unit is 6.5 m wide and 13.1 m long, with a power generation capacity of 18.2 kW. The ROSA solar array uses standard multi-junction GaAs cells installed on a polyimide thin-film substrate. The solar array is rolled and pressed on the central axis in the folding state and is deployed after entering orbit using two compressed composite material rods.

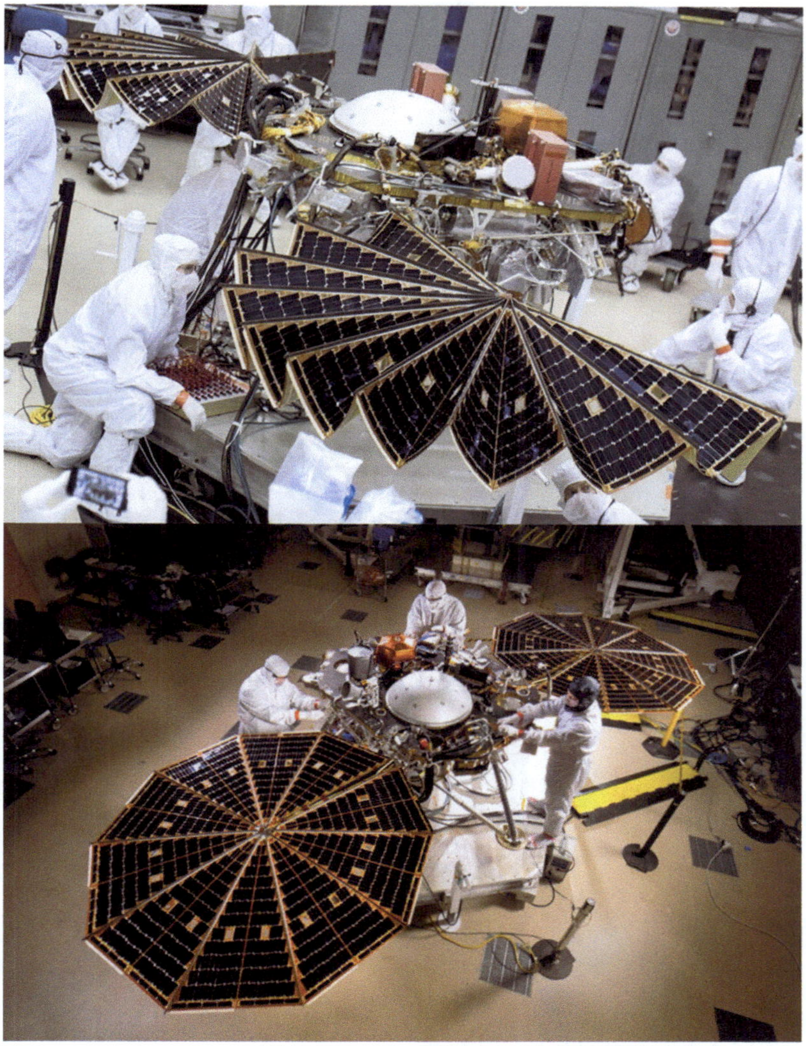

Fig. 4.22 UltraFlex flexible solar array folding and unfolding. (Credit: NASA)

The width of the ROSA solar array unit is limited by the envelope of launcher, and a longer array structure can be achieved in the length direction through rolling, mainly limited by the deployment capability and rigidity of the composite rod. High-power solar arrays can be formed through multiple solar array units, and future ROSA solar arrays can be expanded to hundreds of kilowatts to 1 MW (Table 4.2). The ROSA solar array has been applied to the DART mission in the United States and the solar array update mission of the International Space Station (Fig. 4.24).

4. Two-dimensional folding and unfolding

Two-dimensional folding and unfolding is an important technological direction for achieving larger size flexible solar arrays in the future, especially for solar arrays with a length-to-width ratio close to 1. It can further enhance the mass-to-power ratio and area-to-power ratio, reduce the volume when folded, but compared to the one-dimensional folding and unfolding form, due to the need to consider the impact of the thickness of the solar array on the two-dimensional folding and unfolding, its folding and unfolding methods

Table 4.1 Expansion of MegaFlex solar array

Launch rocket	Delta IV	Falcon 9	Ariane5	SLS PF1B	SLS PF2
Fairing diameter/m	4	5.2	5.4	8.4	10
Solar array diameter/m	20	25	25	30	30
Output power/kW	190	300	300	440	440

Fig. 4.23 Flexible solar array of Hubble Telescope. (Credit: NASA)

Table 4.2 Expansion of ROSA solar array

Parameter	Type I	Type II	Type III
Total power of array/kW	300–400	700	1000
Array unit width/m	7.9	7.9	7.9
Array unit length/m	15.5	19.8	22.4
Array unit power/kW	25–33.33	43.75	50
Number of array units	12	16	20

have greater complexity. At present, several two-dimensional folding and unfolding methods of thin-film solar arrays have been studied.

(a) Edge-to-edge two-dimensional folding and unfolding method. For rectangular thin-film solar arrays, the solar array is divided into multiple rectangular subarrays, with a certain distance left between the subarrays for folding. Z-shaped folding is performed in two directions, as shown in Fig. 4.25. First, alternate folding is performed in the *X* direction, then in the *Y* direction, and finally, it is folded into an area slightly larger than the subarray. The height direction is related to the number of subarrays and is also related to

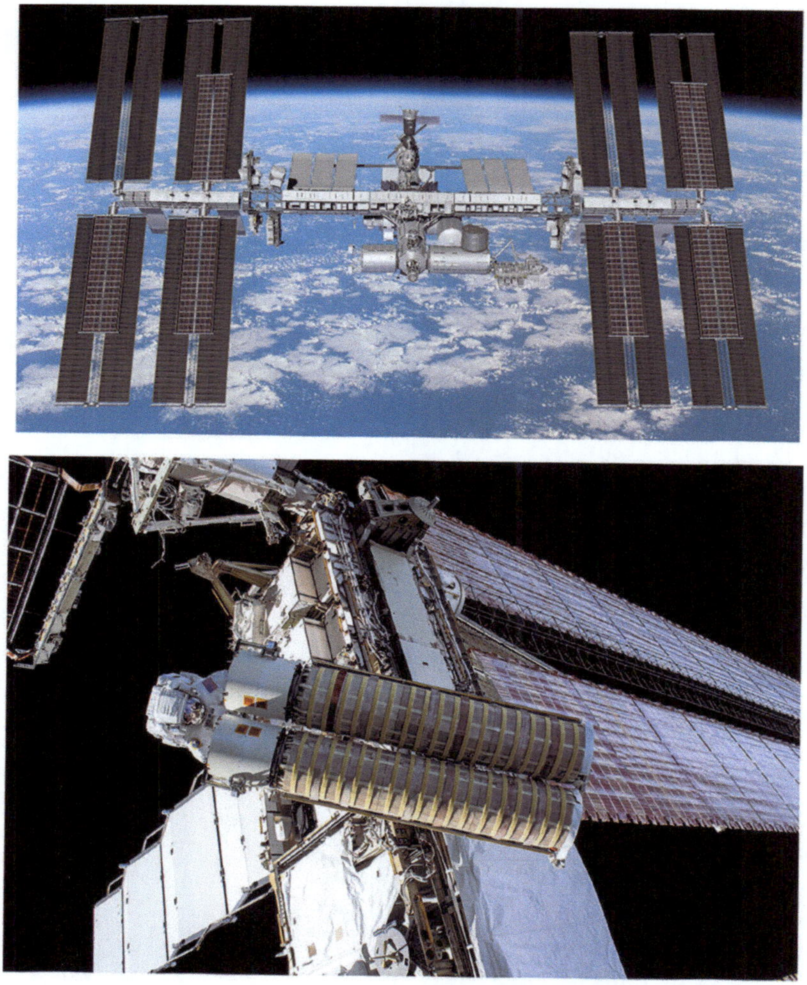

Fig. 4.24 ROSA flexible solar array of International Space Station. (Credit: NASA)

the folding bending radius. The unfolding sequence is the opposite of the folding process, first unfolding in the *Y* direction, then in the *X* direction. The unfolding process of the related thin-film solar array prototype developed is shown in Fig. 4.25. The solar array is unlocked from the fully folded state, controlled to unfold using a truss drive mechanism, and after it is in place, the array surface is tensioned at multiple points through the truss to ensure the flatness of the solar array.

(b) Diagonal two-dimensional folding and unfolding method. For rectangular thin-film solar arrays, a two-dimensional folding method can also be used in the diagonal direction. Figure 4.26 shows a flexible solar array scheme called GoSolAr (Gossamer Solar Array) designed by Germany. This scheme is mainly based on CIGS thin-film cells and carbon fiber pod rod technology. The solar array is divided into multiple rectangular subarrays, with a certain distance left between the subarrays for folding, and Z-shaped folding is performed in two directions. CIGS thin-film cells are pasted on a flexible substrate, arranged in rows in the diagonal direction, then alternately folded towards the center in one diagonal direction, and then alternately folded towards the center in the other diagonal

Fig. 4.25 Two-dimensional unfolding process of thin-film solar array. (Credit: SAST)

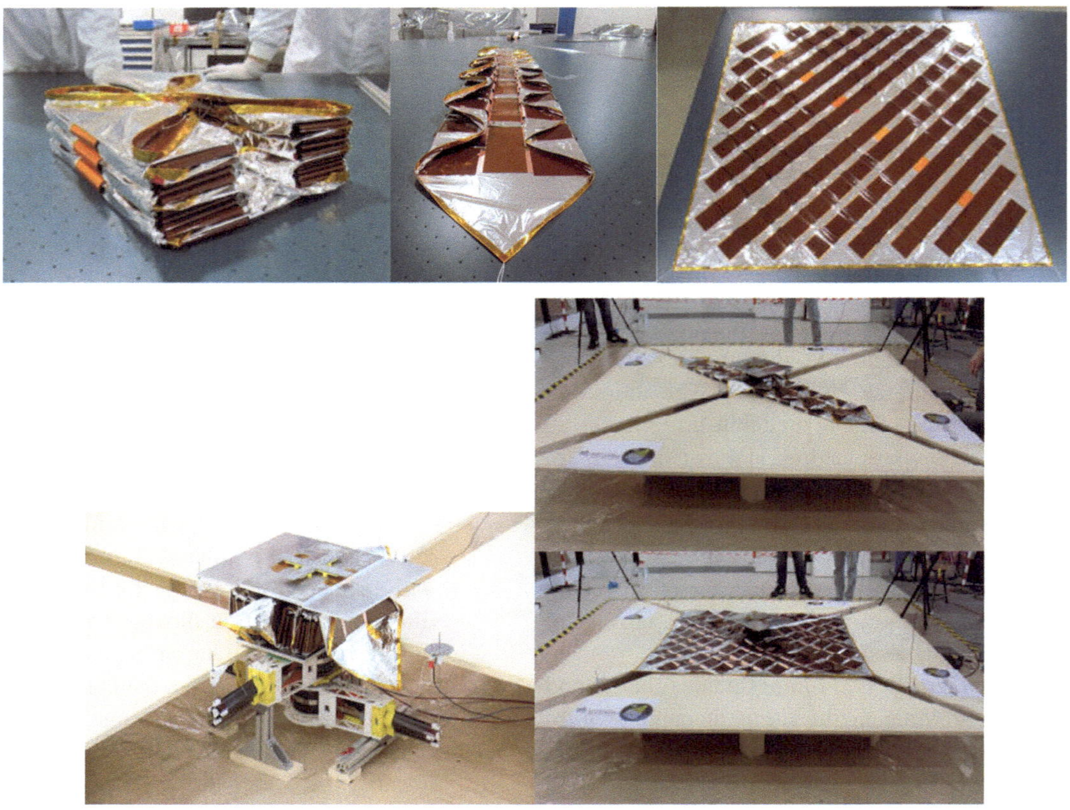

Fig. 4.26 Two-dimensional folding and unfolding process of GoSolAr thin-film solar array. (Credit: DLR)

Fig. 4.27 Two-dimensional folding and unfolding of Caltech thin-film structure. (Credit: Caltech)

direction. Finally, it is folded into an area slightly larger than the four subarrays. The height direction is related to the number of folding layers and also needs to consider the folding bending radius. The entire solar array is supported by two carbon fiber pod rods arranged up and down in the diagonal direction. The two carbon fiber pod rods are wound around the center in the folded state and are unfolded in two directions respectively by motor control during unfolding. The principle prototype and the folding and unfolding process are shown in Fig. 4.26.

(c) Winding two-dimensional folding and unfolding method. In order to greatly reduce the mass of the Solar Power Satellite, Caltech conducted research on the folding and unfolding technology of flexible sandwich structures and flexible thin-film structures, and proposed a winding two-dimensional folding and unfolding scheme similar to a solar sail. As shown in Fig. 4.27, multiple strips are divided from the outside to the inside along the length direction of the four sides of the thin-film structure. Each strip can paste strip-shaped flexible sandwich structure modules, and then the strips are folded from the outside to the inside in a Z shape. The shape is folded to form a cross-shaped strip, which is then wound around a central axis to form a cylindrical structure. The support structure of this

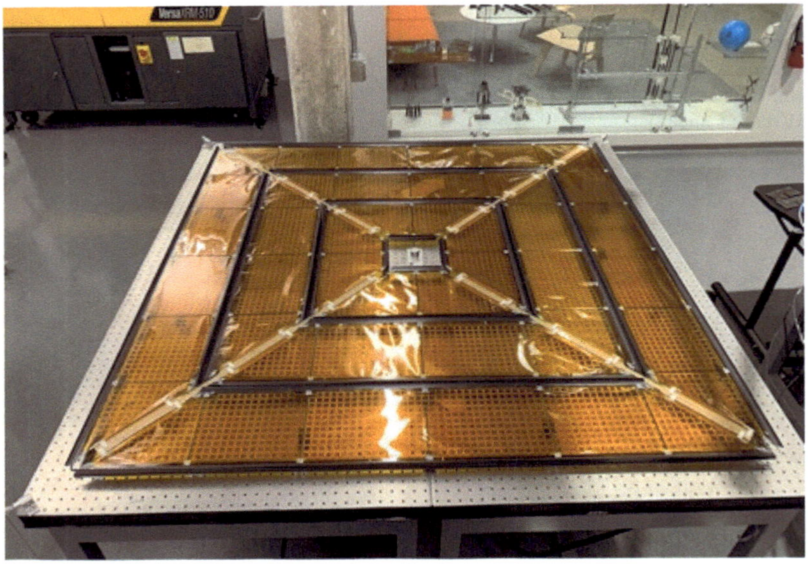

Fig. 4.28 Two-dimensional unfolded state of Caltech thin-film structure. (Credit: Caltech)

scheme is two elastic rods in the diagonal direction, which are wound together with the cross-shaped strip. When unfolding, the elastic rods are controlled by the mechanism to unwind and unfold in the length direction, finally achieving the simultaneous two-dimensional unfolding of the entire solar array (Fig. 4.28).

4.4 Space Power Transmission and Management Method

The power generated by the solar array of SPSs needs to be transmitted to the microwave transmitting antenna and converted into microwaves for transmission. Ideally, the solar array (or concentrator) needs to be sun-pointing, and the transmitting antenna needs to be oriented towards the ground receiving device. Within one orbital period, the relative position between the solar array (or concentrator) and the transmitting antenna changes by 360°. Due to the great technical difficulties in long distance high-power electric power transmission in space, especially the high-power conductive rotary joint for rotating power transmission, how to transmit the power generated by the solar array to the microwave transmitting antenna has become the most important priority in the design of Solar Power Satellite. Space power transmission and management methods can be classified in different ways, such as whether to use conductive rotary joint, high-voltage or low-voltage power transmission, short-distance or long-distance power transmission, DC or AC power transmission, and centralized or distributed power supply. The following focuses on the analysis from the perspective of conductive rotary joints.

1. Single conductive rotary joint type

   The use of a conductive rotary joint means that the solar array and the microwave transmitting antenna are in a separated state, and it can ensure the sun-pointing of the solar array and the ground-pointing of the transmitting antenna, thus ensuring stable power generation and microwave power transmission. It requires the configuration of a very high-power conductive rotary joint and the use of high-voltage long distance power transmission. Representative concepts include the 1979 SPS reference system, SSPS-OMEGA, etc. (Fig. 4.29). For the single conductive rotary joint type transmission method, the GW-level conductive rotary joint meets great technical challenges.

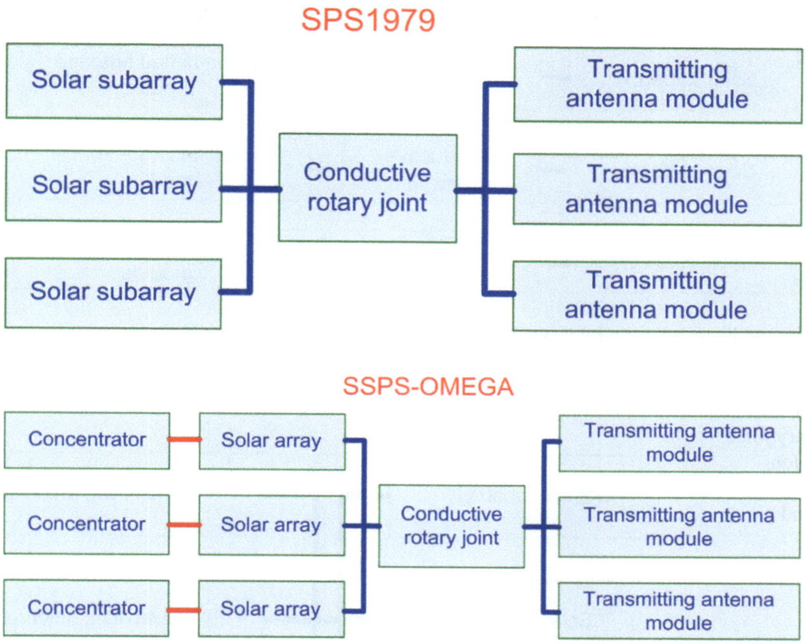

Fig. 4.29 Single conductive rotary joint space power transmission and management

2. Multiple conductive rotary joint type
   In order to ensure the sun-pointing of the solar array and the ground-pointing of the transmitting antenna, and at the same time to reduce the technical difficulty of the GW-level conductive rotary joint, one solution is to use multiple conductive rotary joints for parallel transmission to reduce the transmission power. However, due to the special nature of spacecraft on-orbit operation, a special configuration needs to be designed to decompose the integrated solar array into multiple independently rotating solar subarrays, each of which transmits power through its own conductive rotary joint. A representative concept is the MR-SPS (Fig. 4.30), which reduces the GW-level conductive rotary joint to the 10 MW level, greatly reducing its technical difficulty. The difficulty of the multiple conductive rotary joint type is that it still needs to use high-voltage long distance power transmission.
3. Non-conductive rotary joint type
   The absence of a conductive rotary joint means that there is no relative motion between the solar array and the microwave transmitting antenna. It is impossible to achieve solar array orientation towards the sun and antenna orientation towards the Earth; therefore, it is impossible to ensure stable power generation and microwave power transmission. Typical representative concepts include the SunTower SPS and Tethered-SPS (Fig. 4.31). The SunTower SPS uses gravity gradient attitude stabilization, the transmitting antenna maintains Earth orientation, but the solar array cannot achieve Sun orientation, so power generation is discontinuous. The corresponding power transmission method is shown in Fig. 4.31, and high-voltage long distance power transmission is still required. The Tethered-SPS uses a special sandwich structure. Since the solar array and the microwave transmitting antenna are located on both sides of the sandwich struc-

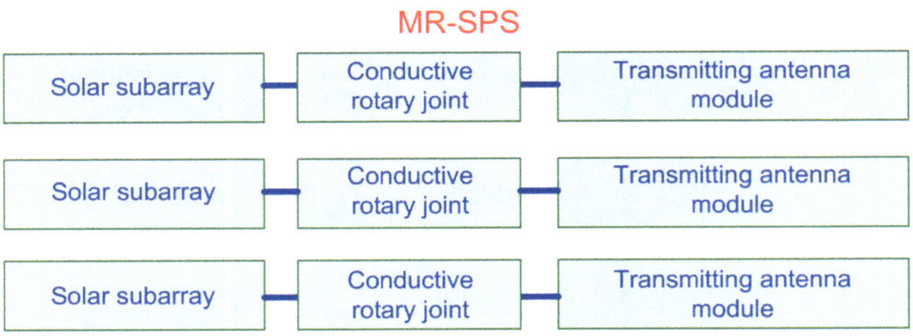

Fig. 4.30 Multi-conductor rotary joint space power transmission and management

Fig. 4.31 Non-conductive rotary joint space power (Non-concentration) transmission and management

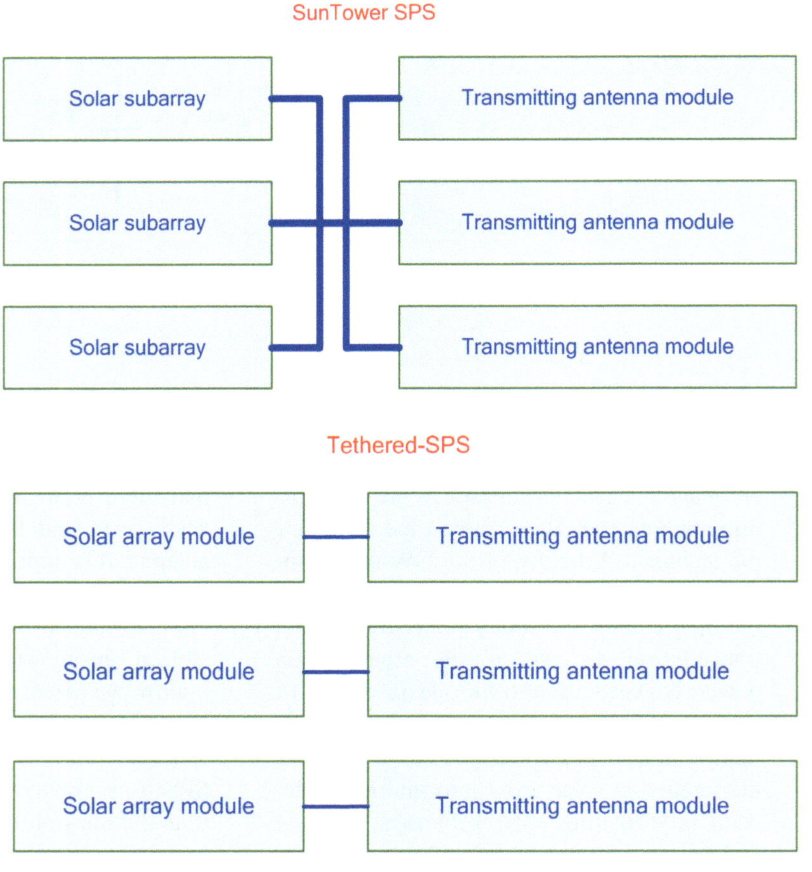

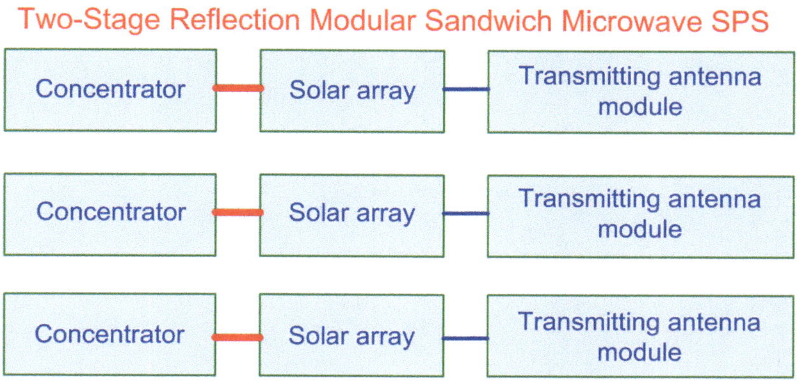

Fig. 4.32 Non-conductive rotary joint (Concentration) space power transmission and management

ture, while maintaining the microwave transmitting antenna towards the Earth, the illumination of the solar array will maintain periodic changes, so power generation is discontinuous. The advantage of using a sandwich structure is that it eliminates the need for long distance power transmission, greatly simplifying the difficulty of power transmission.

In the absence of conductive rotary joints, if you want to achieve relatively stable power generation and power transmission within the orbit period, you need a special design. One method is to use a sun-pointing concentration system, a typical representative concept is the Two-Stage Reflection Modular Sandwich Microwave SPS (Fig. 4.32). This concept uses a concentration system including a main reflector and a secondary reflector. The main reflector can track the sun within the orbit period and focus the incident light to the secondary reflector. The secondary reflector then reflects the sunlight to the surface of the solar cell in the sandwich structure, achieving relatively stable power generation under the condition of the microwave transmitting antenna towards the Earth. This scheme simplifies the difficulty of power transmission due to the use of a sandwich structure, but how to ensure a relatively uniform concentration illumination distribution within the entire orbit position and the heat dissipation of the sandwich structure under high concentration ratio are two biggest technical challenges.

Bibliography

1. Grundmanna J T, Spietz P, Seefeldt P, et al. GOSSAMER Deployment Systems for Flexible Photovoltaics[C]. 67th International Astronautical Congress, Guadalajara, Mexico, 2016.
2. Gdoutos E E, Leclerc C, Royer F. A Lightweight Tile Structure Integrating Photovoltaic Conversion and RF Power Transfer for Space solar power Applications[C]. 2018 AIAA Spacecraft Structures Conference, Kissimmee, USA, 2018.
3. Pedivellano A, Gdoutos E E, Pellegrino S. Sequentially Controlled Dynamic Deployment of Ultra- Thin Shell Structures[C]. SciTech 2020, Orlando, USA, AIAA-2020-0690.
4. Gdoutos E E, Truong A, Pedivellano A, et al. Ultralight Deployable Space Structure Prototype[C]. SciTech 2020, Orlando, USA, AIAA-2020-0692.
5. Meng X L, Xia X L, Sun C, et al. Optimal Design of Symmetrical Two-Stage Flat Reflected Concentrator[J]. Solar Energy, 2013, 93: 334-344.
6. Hou X, Wang L, Liu Z. High-Voltage and High-Power Electricity Generation, Transmission and Management of MR-SPS[J]. Advances in Astronautics Science and Technology, 2022, 5 (1): 31-38.
7. Wu M, Liu Q, Qian M. Fabrication and Experimental Investigation of Flexible Thin Film Solar Module with Ultra-high Voltage for the Space Power Satellites[J]. Advances in Astronautics Science and Technology, 2022, 5 (1): 59-63.
8. Chen Q, LiuZG, Zhang X F, et al. Spacecraft Power System Technology[M] (in Chinese). Beijing: Beijing Institute of Technology Press, 2018.
9. Duan Z Z. Optical Analysis and Optimization of the Concentrator of SSPS-OMEGA[D](in Chinese). Xidian University, 2017.6.
10. Fan G H. Heat Dissipation Analysis and Innovative Design of Sunlight Concentration and Photoelectric Conversion System for Space Solar Power Station[D] (in Chinese). Xidian University, 2022.4.
11. https://images.nasa.gov/

5 Wireless Power Transmission System

5.1 Microwave Power Transmission

5.1.1 Composition and Characteristics of the Microwave Power Transmission System

The microwave power transmission system of SSP transmits energy to the Earth wireless in the form of microwaves, and the ground receiving system converts the microwave power into electric power, which needs to go through the DC-RF, microwave transmission, and RF-DC three main processes.

Microwave power transmission includes space and ground segments. The space segment is mainly used to efficiently convert electricity into microwaves, and use large-scale transmitting antennas for high-precision transmission to the ground. The ground segment receives the microwave beam as much as possible, efficiently converts it into electricity and connects it to the power grid (Fig. 5.1). In order to achieve high-precision microwave power transmission, it also includes pilot beam emitting on the ground and pilot beam reception and processing in space. The design of microwave power transmission system mainly focuses on microwave frequency, system efficiency, beam accuracy, and system scale.

Microwave power transmission is similar to microwave wireless communication, and the main differences include the following aspects.

1. For wireless power transmission, microwave is the carrier of energy, which are directly rectified into direct current at the receiving end to supply the load, so the emitted microwave is a point frequency RF. The microwave in wireless communication is the carrier of information. In order to achieve a higher communication capacity, it is necessary to transmit wideband microwaves and demodulate the information carried by the microwaves at the receiving end.
2. Microwave power transmission focuses on the efficiency of power transmission, requiring high microwave interception efficiency. Therefore, the transmitting antenna should be large enough to generate a narrow beam, and the receiving rectenna should be large enough to cover the main beam range (Fig. 5.2). At the same time, efficient microwave transmitter and microwave rectifier are needed, and the accuracy of beam pointing is high. Microwave wireless communication is more concerned with the range of wireless communication and the quality of the received signal. The transmitted beam is generally a wide beam, and the receiving end only needs to receive a microwave signal that meets the signal-to-noise ratio requirement. Therefore, the sizes of the transmitting antenna and the

X. Hou, *Space Solar Power*, Sustainable Development Goals Series,
https://doi.org/10.1007/978-981-97-3580-8_5

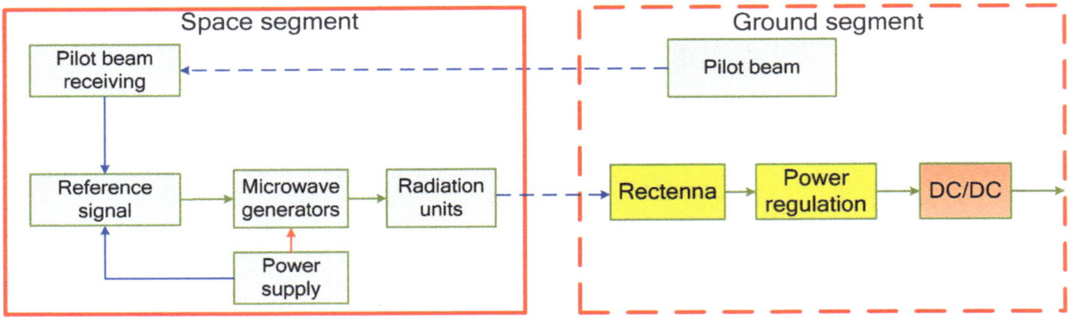

Fig. 5.1 Diagram of microwave power transmission

Fig. 5.2 Difference in beam distribution between microwave power transmission and microwave wireless communication

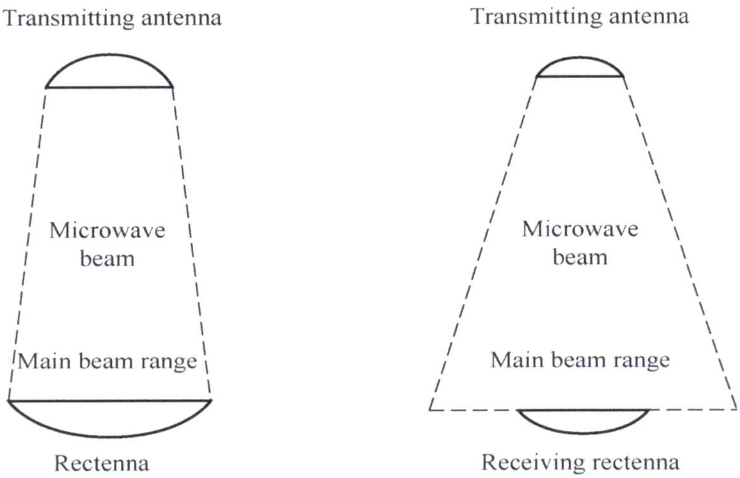

receiving antenna are smaller, and the accuracy of beam pointing is lower.

3. The power of wireless power transmission is large, and the corresponding microwave power intensity at the receiving end is high. The power transmitted by wireless communication is small, and the corresponding microwave power intensity at the receiving end is low, which is several orders of magnitude lower than that of microwave power transmission.

5.1.2 Efficiency Chain of the Microwave Power Transmission System

The efficiency chain of the microwave power transmission system is shown in Fig. 5.3.

The main influencing factors include DC-RF conversion efficiency, beam collection efficiency (BCE), and RF-DC conversion efficiency. Among them, DC-RF conversion efficiency and RF-DC conversion efficiency are related to devices and circuits, while BCE is related to microwave frequency, transmitting antenna size, receiving antenna size, and transmission distance.

5.1.3 Microwave Frequency Choice

The Solar Power Satellite transmits energy from space to the ground, requiring the energy loss of microwaves passing through the atmosphere to be as small as possible. The factors affecting the microwave transmittance mainly include atmospheric molecules (oxygen, water vapor, etc.), weather (including rain, fog, snow, hail, clouds,

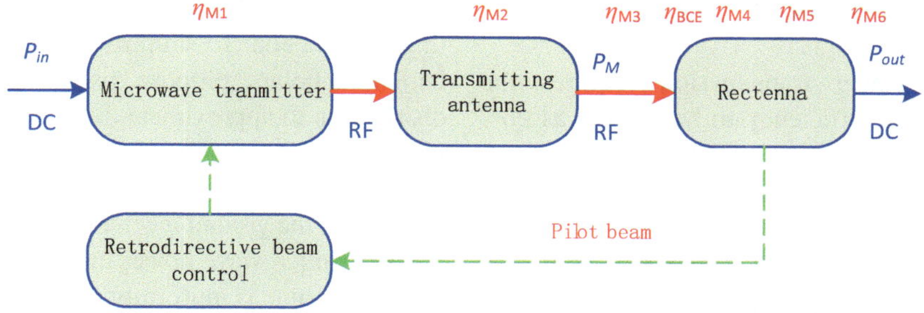

Fig. 5.3 Efficiency chain of microwave power transmission system. P_{in}—Input electric power, P_M—Transmitting microwave power, P_{out}—Electric power output of the rectenna, η_{M1}—DC-RF efficiency, η_{M2}—Antenna radiation efficiency, η_{M3}—Microwave transmission efficiency, η_{M4}—Beam collection efficiency (BCE), α_M—Microwave absorptivity of the rectenna, η_{M5}—RF-DC efficiency, η_{M6}—Electrical regulation efficiency

Fig. 5.4 Atmospheric attenuation of electromagnetic waves at different frequencies

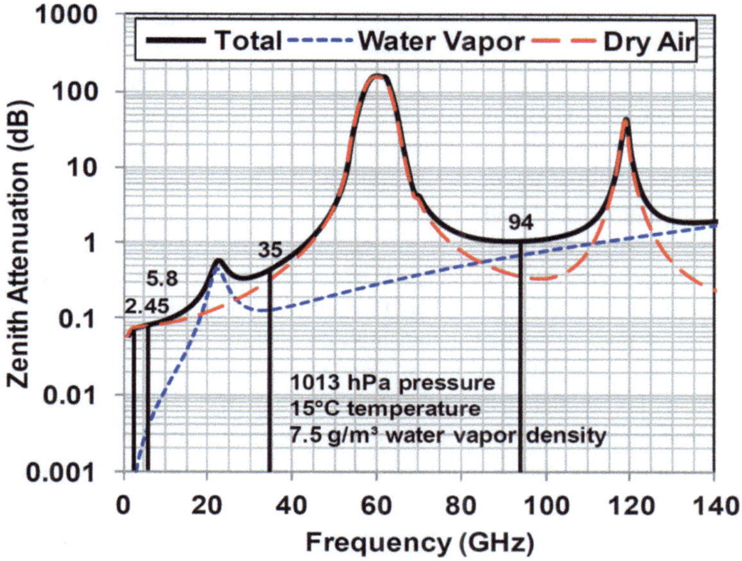

etc.), and suspended particles in the atmosphere (dust, smoke, etc.). According to the attenuation curve of electromagnetic waves of different frequencies (Fig. 5.4), the microwave frequency passing through the atmosphere should be less than 10 GHz. At the same time, the choice of frequency should also consider the efficiency of the microwave transmitter, rectification efficiency, ITU's allocation for this frequency, etc. and should comprehensively consider the size matching of the transmitting antenna and receiving antenna and the restrictions on the microwave power intensity of the transmitting and receiving.

At present, the research on SSP internationally generally chooses the Industrial, Scientific, and Medical (ISM) frequency bands of 2.45 and 5.8 GHz, and some researches choose 10 GHz. The atmospheric transmittance of 2.45 GHz microwaves is better than that of 5.8 GHz, but the corresponding transmitting antenna is about twice as large. The atmospheric transmittance of 10 GHz microwaves is slightly lower than that of 5.8 GHz, but the corresponding transmitting antenna is about half of that of 5.8 GHz. Considering the development of technology and the size of the antenna, the current GW-level SPS usually chooses the microwave frequency of 5.8 GHz, and the corresponding atmospheric attenuation rate does not exceed 5%.

5.1.4 Antenna Sizes

Microwave power transmission requires the beam collection efficiency to be as high as possible. According to the experimental results of Goubau and Brown, the beam collection efficiency depends on the microwave wavelength, power transmission distance, and the diameter of the transmitting and receiving antennas. The impact of these factors on beam collection efficiency can be characterized by the parameter τ:

$$\tau = \frac{\sqrt{A_t A_r}}{\lambda D} \tag{5.1}$$

where

A_t—Area of transmitting antenna
A_r—Area of the receiving antenna
λ—Wavelength of the microwave beam
D—Transmission distance

The relationship between τ and beam collection efficiency is shown in Fig. 5.5. When τ is large enough (such as $\tau > 2$), the beam collection efficiency of microwave power transmission is close to 100%, provided that the microwave power intensity is Gaussian distributed on the aperture of the transmitting antenna (Fig. 5.6). The Gaussian distribution can cause significant difficulties in power supply and heat dissipation at the transmitting antenna end. To simplify the Gaussian distribution, scholars propose using a step-shaped distribution to approximate the Gaussian distribution (Fig. 5.6), resulting in the ground receiving beam distribution shown in Fig. 5.7. Figure 5.7 also shows the ground receiving beam distribution when the microwave power intensity of the transmitting antenna is uniformly distributed. It can be seen that when the microwave is uniformly distributed at the transmitting end, the corresponding ground receiving beam is sharper, that is, the center power intensity is higher, the main beam is narrower, but the side lobe power intensity is also higher, which will lose more energy. Using a step distribution will be closer to the results obtained by the Gaussian distribution, the center power intensity is reduced, the main beam is wider, but the side lobe power intensity is reduced, and the capture efficiency is improved.

The size of the transmitting and receiving antennas of SSP needs to consider factors such as microwave frequency, transmission distance, and microwave power intensity restrictions on the ground. When the microwave frequency and transmission distance are determined, the diameters of the transmitting antenna and the receiving antenna are inversely proportional. For a 5.8 GHz microwave, the wavelength is 5.17 cm; the

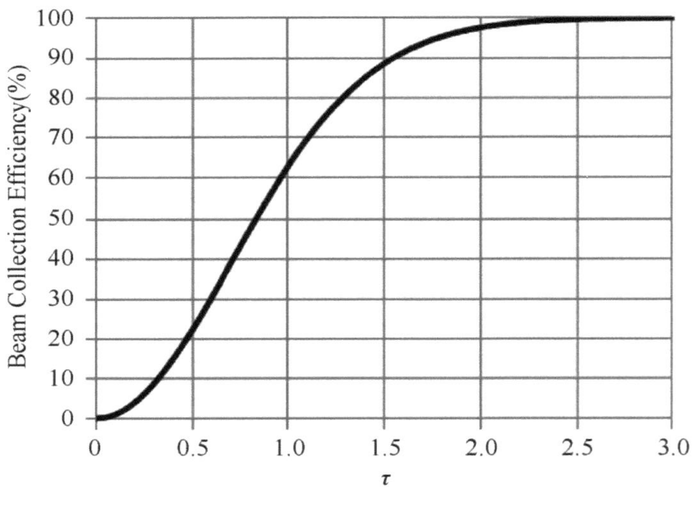

Fig. 5.5 Relationship between BCE and τ

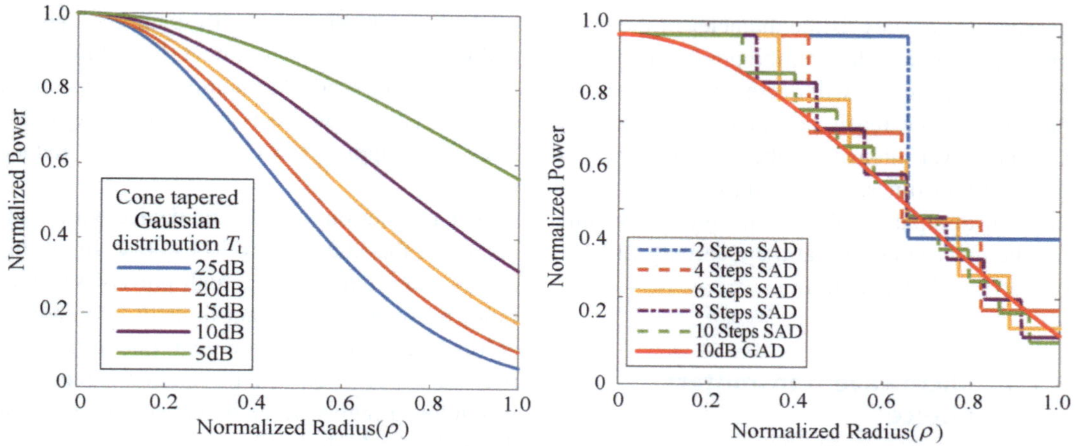

Fig. 5.6 Normalized power intensity distribution of transmitting antenna with different cone tapered Gaussian distribution and different step distribution

Fig. 5.7 Normalized power intensity distribution at the receiving end under uniform distribution and different step distribution of transmitting antenna

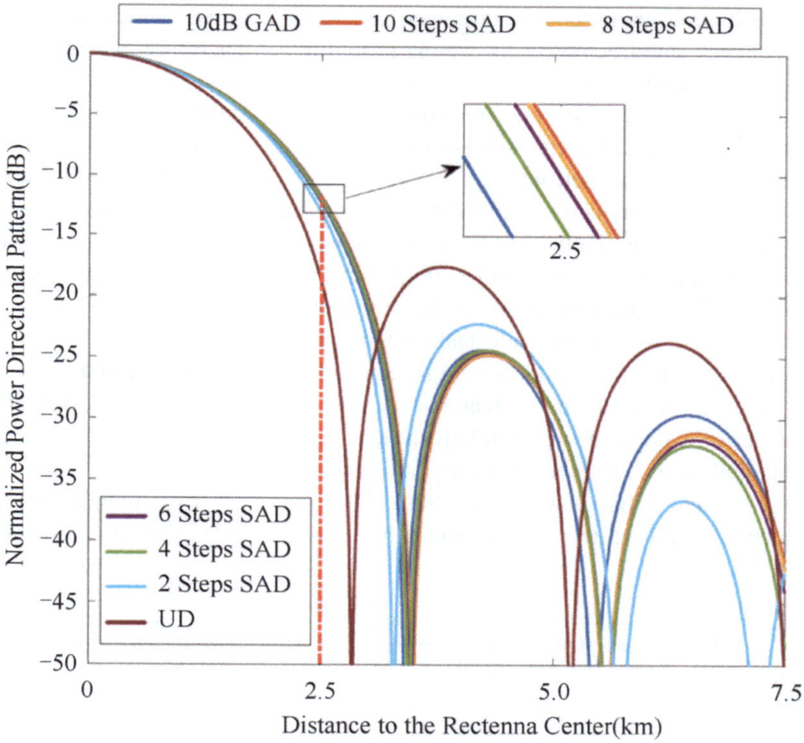

geosynchronous orbit altitude is 35,786 km to achieve 95% beam collection efficiency; the corresponding τ is about 2. According to the typical design of an SPS, the corresponding transmitting antenna diameter is about 1 km, and the receiving antenna diameter is about 5 km. If the diameter of the transmitting antenna is reduced by half, the power intensity at the transmitting end will increase four times, the diameter of the receiving antenna will be double, and the power intensity at the receiving end will decrease four times, requiring the transmitting antenna to have better heat dissipation capacity. If the diameter of the transmitting antenna is double, the power intensity at the transmitting end will decrease four times, the diameter of the receiving antenna will be half, and the power intensity at the receiving end will increase four times, which is more beneficial for the heat dissipation of the transmitting antenna, but the power intensity at the receiving end may exceed the allowable limit.

5.1.5 Microwave Transmitters Choice

The microwave frequency emitted by the SPS is a point frequency, and bandwidth does not need to be considered. The choice of the microwave transmitters mainly considers power, efficiency, frequency stability, power-to- weight ratio, lifespan, adaptability to space environment, power supply, cost, etc. Currently, the main considerations are semiconductor solid-state microwave amplifiers and magnetrons.

The semiconductor solid-state microwave power source is a microwave amplifier, which can obtain stable microwave frequency and phase. The current main development is power devices represented by wide bandgap materials such as gallium nitride (GaN) and silicon carbide (SiC). The main technical features include: the power of a single device is relatively low, at the hundred-watt level; the efficiency is increasing rapidly, and it has now reached 80%; high frequency stability; the power-to-weight ratio is gradually increasing; long lifespan; good adaptability to the space environment (including radiation tolerance and high-temperature operation characteristics); the cost is relatively high, but large-scale application can significantly reduce the cost.

The magnetron is a vacuum device with high-power, at the kilowatt level or above; high efficiency, can reach 80%; high-power-to-weight ratio; lifespan is somewhat limited; good adaptability to the space environment; due to the centralized power, good heat dissipation is required; it requires a high voltage power supply of about 5000 V, with high requirements for the stability of the power supply voltage; the cost is relatively low. The biggest difficulty in the application of magnetrons is their poor frequency and phase stability, which can undergo random changes. Therefore, in order to achieve efficient coherent power synthesis, it is necessary to use frequency stabilization and phase locking technology to develop phase control magnetrons (PCM). Figure 5.8 shows a method of frequency stabilization and phase locking for a phase control magnetron. The system is relatively complex and will have some impact on efficiency.

5.1.6 Microwave Power Reception and Conversion

The reception and conversion of microwave energy is mainly achieved through the receiving antenna array and rectifying circuit. The micro-

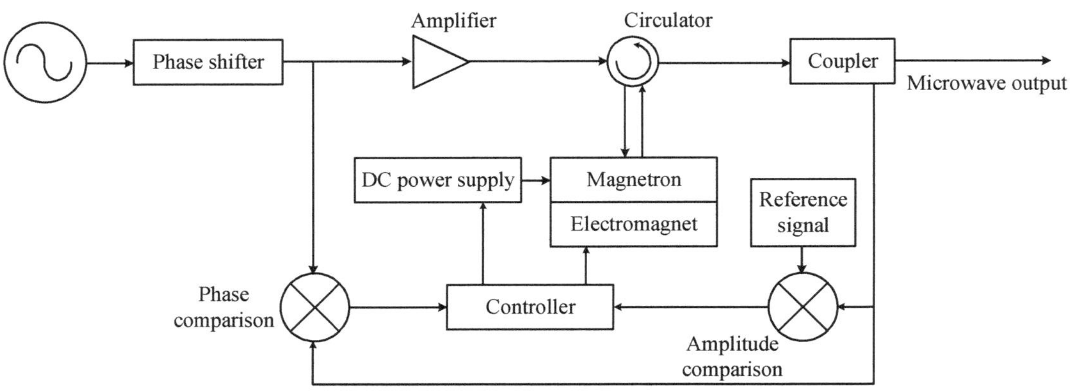

Fig. 5.8 Phase locking method of phase controlled magnetron. (Credit: Sichuan University)

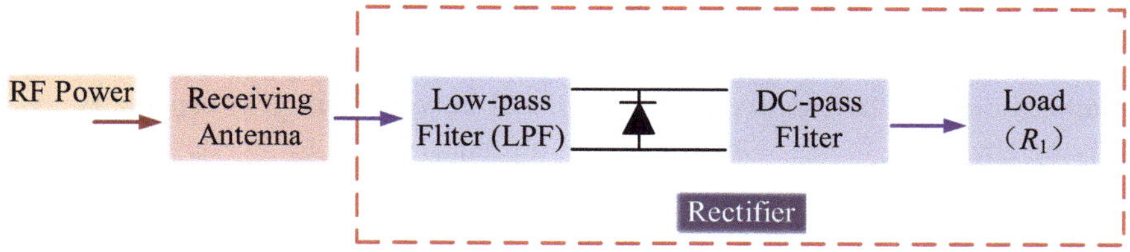

Fig. 5.9 Rectifying circuit structure diagram

Fig. 5.10 Typical rectifying circuit. (Credit: Sichuan University)

wave energy is received by the receiving antenna and converted into direct current by the rectifying circuit. It is then output through direct current series and parallel combination, voltage transformation or inversion, etc.

Typical rectifying circuits include low-pass filters, rectifying diodes, bypass filters, and loads (Figs. 5.9 and 5.10). The main functions of the low-pass filter include: only allowing the fundamental frequency microwave to pass, preventing other frequency components from entering the rectifying circuit; reflecting the high-order harmonics generated by the diode; achieving impedance matching between the rectenna and the filter. The function of the bypass filter is to only allow direct current to pass, reflecting the residual fundamental frequency and harmonics above the fundamental frequency back to the rectifying diode, improving the smoothness and energy utilization rate of the output direct current. Microwave nonlinear devices are the core of the microwave rectifying circuit, usually choosing Schottky diodes. The rectification efficiency of the rectifying circuit is affected by the characteristics of the microwave nonlinear device, as well as external factors such as microwave injection power, impedance matching, and load. An ideal diode combined with an appropriate rectifying circuit design can achieve high rectification efficiency. The diode has its own junction voltage and breakdown voltage. If the input voltage of the diode is lower than the junction voltage or higher than the breakdown voltage, the diode cannot work in the best state. In order to make the diode achieve high work efficiency, an appropriate matching circuit is needed to complete the impedance transformation, so as to ensure that the rectifying circuit can reach resonance and be in the best working state.

Due to the large difference in power intensity distribution of the ground receiving antenna, considering the optimal operation power range of the rectifying circuit, a unified rectifying circuit method cannot be adopted. Therefore, it is necessary to divide the receiving antenna into zones according to the microwave power intensity and the dynamic range of power intensity changes caused by beam deviation. For moderate power intensity, a single diode rectifying circuit corresponding to a single antenna unit is used. For antenna units with larger power, considering the tolerance of the diode, a rectifying circuit designed with multiple diodes in series and parallel can be used. For antenna units with lower power, it is considered to use the method of power synthesis of received microwaves from multiple antenna units and then rectify the output with a rectifying circuit.

5.1.7 Microwave Transmitting Antenna

Microwave energy emission generally uses parabolic antennas or microstrip array antennas (Fig. 5.11). The parabolic antenna uses a feed source located at the focus of the parabola to radiate spherical waves, which are then reflected by the parabola to form collimated array waves. Therefore, it has advantages such as good directionality, large power capacity, high gain, and high radiation efficiency. Due to the huge size (kilometer-scale) of the microwave transmitting antenna of the SPS, it is impossible to use a single parabolic antenna. The structure of multiple parabolic antennas will be very complex, especially the parabolic antenna needs to be deployed in orbit. The technology is complex, and it is difficult to maintain the shape. The microstrip array antenna is composed of multiple microstrip radiating patch units through arraying, and the power distribution feeding network is used to feed each unit in the array. It is lightweight, has a low profile, and can achieve high gain through arraying. Since it is a plane antenna, it is easy to deploy in orbit and easy to assemble into a large array antenna.

For kilometer-scale microwave transmitting antennas, modular design is necessary, from basic antenna units to antenna subarrays to antenna structure modules, antenna structure subarrays, and finally expanded to the entire antenna, as shown in Fig. 5.12. Figure 5.13 shows

Fig. 5.11 Parabolic antenna and microstrip array antenna. (Credit: NRL/Sichuan University)

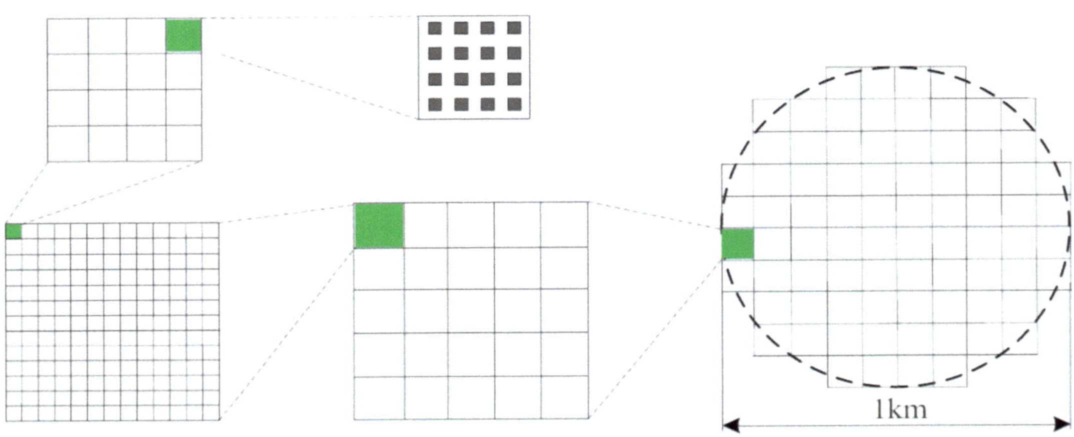

Fig. 5.12 Hierachic modular composition of kilometer-level microwave transmitting antenna

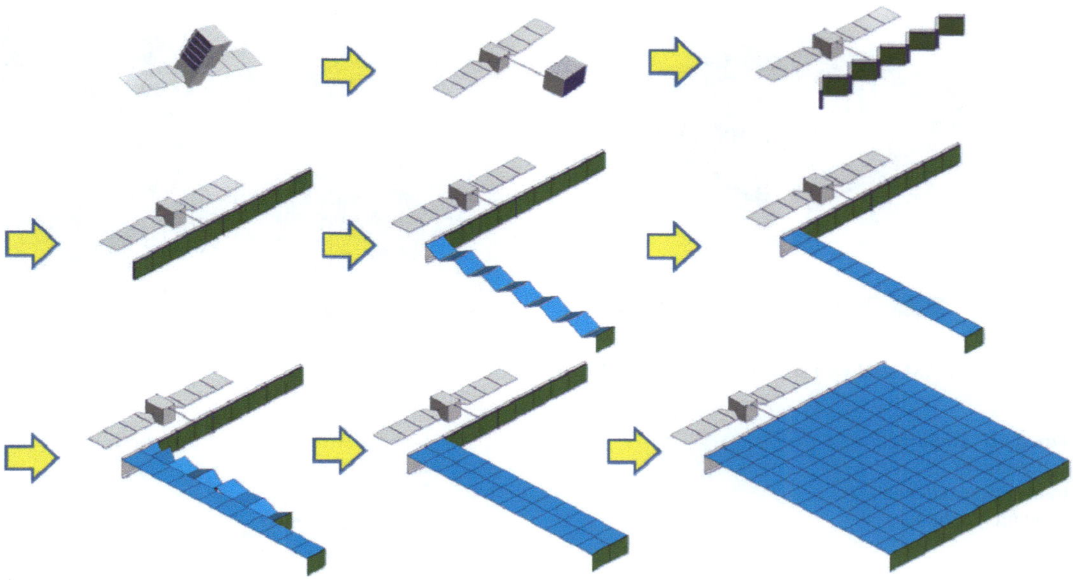

Fig. 5.13 Schematic diagram of large plane antenna deployment in orbit. (Credit: JAXA)

Fig. 5.14 Cross-sectional and energy schematic of sandwich structure

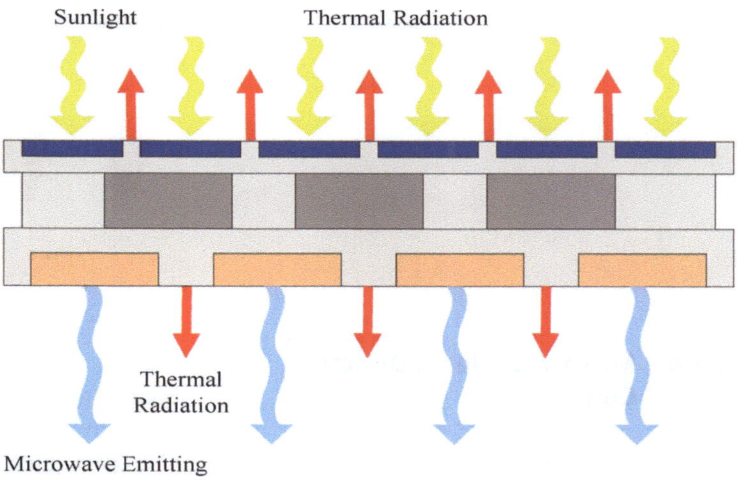

a schematic diagram of the deployment method of a large plane antenna in orbit.

For the sandwich structure SPS, the solar array and the microwave transmitting antenna are integrated into a panel structure, and the structural cross-section is shown in Fig. 5.14. The upper layer is the solar cell, the middle is the sandwich structure, and the lower layer is the transmitting antenna. The sandwich structure is equipped with power regulation equipment, cables, solid-state microwave amplifiers, power dividers, and heat pipes for thermal control. The sunlight is incident from the upper surface, the microwave is emitted from the lower surface, and the heat is mainly radiated from the upper and lower surfaces. Figure 5.15 shows the PRAM test device developed by the US Naval Research Laboratory for on-orbit verification.

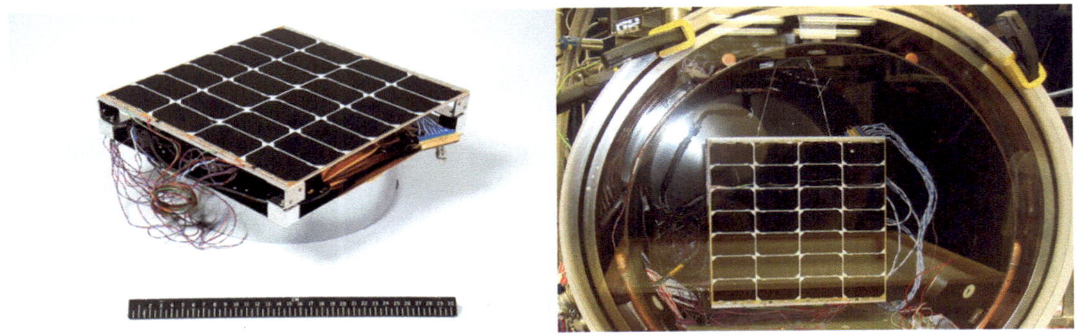

Fig. 5.15 PRAM sandwich structure test device. (Credit: NRL)

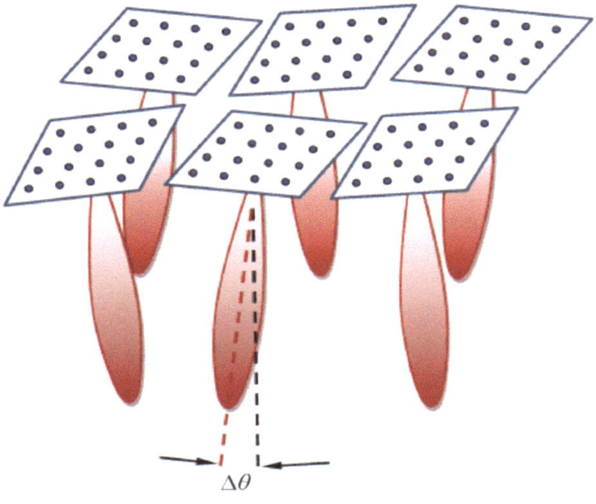

Fig. 5.16 Deformation schematic of transmitting antenna structure module

5.1.8 Microwave Beam Direction Control

The SPS operates in GEO, and the power transmission distance exceeds 36,000 km. In order to achieve efficient energy reception and consider power transmission safety, the accuracy requirements for the microwave beam are extremely high. Assuming that the deviation of the microwave beam center does not exceed 5% of the receiving antenna diameter, for a 5 km diameter receiving antenna, the corresponding beam pointing accuracy should be higher than 0.0004°. Due to the large size and weight of the microwave energy transmitting antenna, it is unrealistic in engineering to achieve a pointing accuracy higher than 0.0004° by mechanical control. At the same time, the microwave energy transmitting antenna needs to be assembled in orbit by many antenna structure modules. Considering the effects of forces, heat, and other space environments, various deformations will occur between the antenna structure modules, including the attitude changes of the antenna structure modules and the position changes between the antenna structure modules, which will greatly affect the efficiency of power synthesis and cause a large change in the overall beam direction (Fig. 5.16). Therefore, it is necessary to steer the beam electronically by adjusting the phase of the antenna unit to achieve high-precision beam pointing. Currently, the main consideration is to use retrodirective beam control technology, that is, the ground receiving sta-

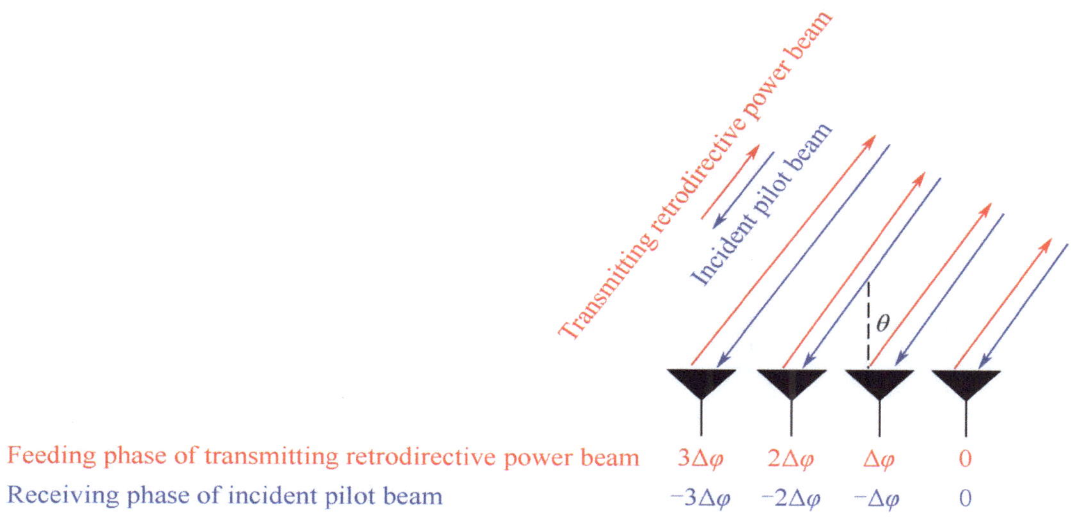

Fig. 5.17 Schematic diagram of phase control principle based on pilot beam

tion sends a pilot beam to the SPS, and the antenna array of the SPS receives the pilot beam to determine the reference phase of different antenna units, and adjust the feeding phase of each antenna unit, so as to accurately transmit the beam to the ground receiving antenna.

Retrodirective beam control is mainly based on the principle of phased array antennas, as shown in Fig. 5.17. When a plane electromagnetic wave is incident on the array antenna, due to the different path lengths of the incident wave reaching each emitting unit, the microwave phase received by each unit lags $\Delta\varphi$ from left to right. There is a linear relationship between phase difference and antenna position:

$$\Delta\varphi = \frac{2\pi f d \sin\theta}{c} = \frac{2\pi d \sin\theta}{\lambda} \quad (5.2)$$

where

$\Delta\varphi$—is the phase difference of the microwaves received between two adjacent units.
d—is the distance between two adjacent units.
f—is the frequency of the incident electromagnetic wave.
θ—is the angle between the direction of the incoming wave and the normal direction of the array.
c—is the speed of light in a vacuum.
λ—is the wavelength.

When the array antenna emits microwaves in the direction of θ, the relationship of phase leading and lagging is exactly the opposite of the receiving situation. The microwave emitted by each unit of the antenna lead by $\Delta\varphi$ from left to right, thereby achieving the maximum power transmission in this direction. According to the previous analysis, retrodirective beam control needs to solve two levels of problems: one is to compensate for the deviation of the overall beam direction of the antenna caused by the position changes between the antenna structure modules; the second is to compensate for the deviation of the beam direction of the antenna structure module caused by the attitude changes of the antenna structure module itself. For the compensation of the overall beam direction deviation of the antenna caused by the position changes between the antenna structure modules, the key is to determine the reference phase of each antenna structure module, so as to ensure that the beams of all antenna structure modules can achieve effective power synthesis. Figure 5.18 provides a phase detection and conjugate circuit implementation method based on pilot beam reception.

For the compensation of the beam direction deviation caused by the attitude change of the antenna structure module, the key is to determine the direction and amount of the attitude change of

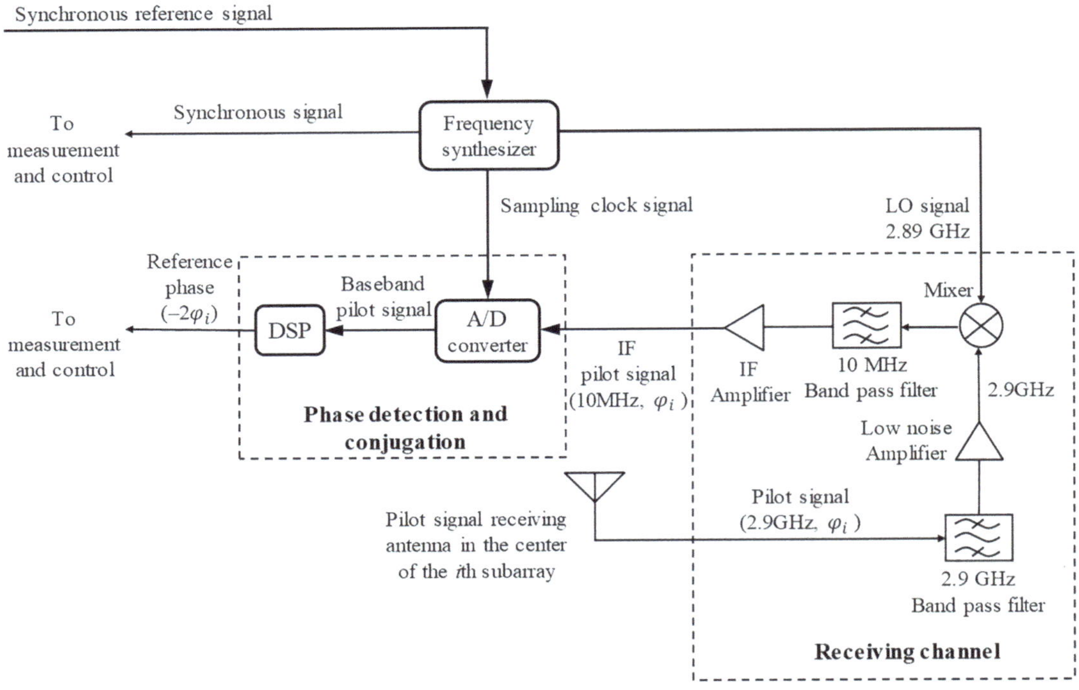

Fig. 5.18 Schematic diagram of phase detection and conjugate circuit method based on pilot beam reception

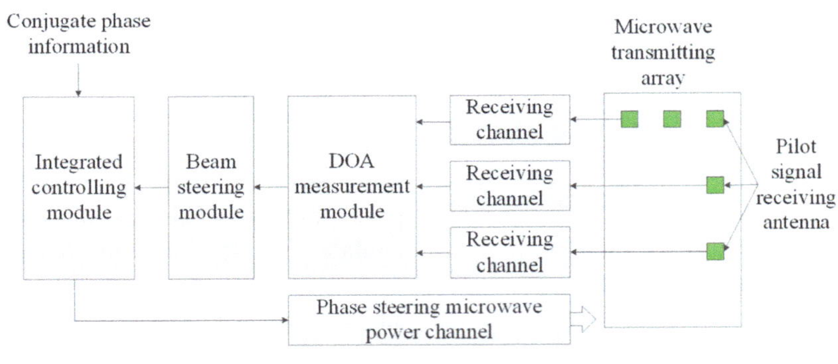

Fig. 5.19 Schematic diagram of direction measurement and phase control method based on pilot beam

the antenna structure module, and then adjust the phase of each related antenna unit in the antenna structure module, so that the emission beam of the antenna structure module automatically corrects its attitude deviation and points to the direction of the incoming beam. Figure 5.19 provides a method for measuring the direction of the incoming beam and controlling the phase based on the pilot beam. Multiple antennas measuring direction of arrival (DOA) are installed in the two-dimensional direction of each antenna structure module, and the direction of the incoming beam is calculated based on the information obtained by the measurement antenna. Based on these parameters, the feeding phase of each antenna unit is analyzed and the phase is controlled, so that the emitting beam points to the direction of the incoming beam.

The entire retrodirective beam control process is shown in Fig. 5.20. The pilot beam is emitted from

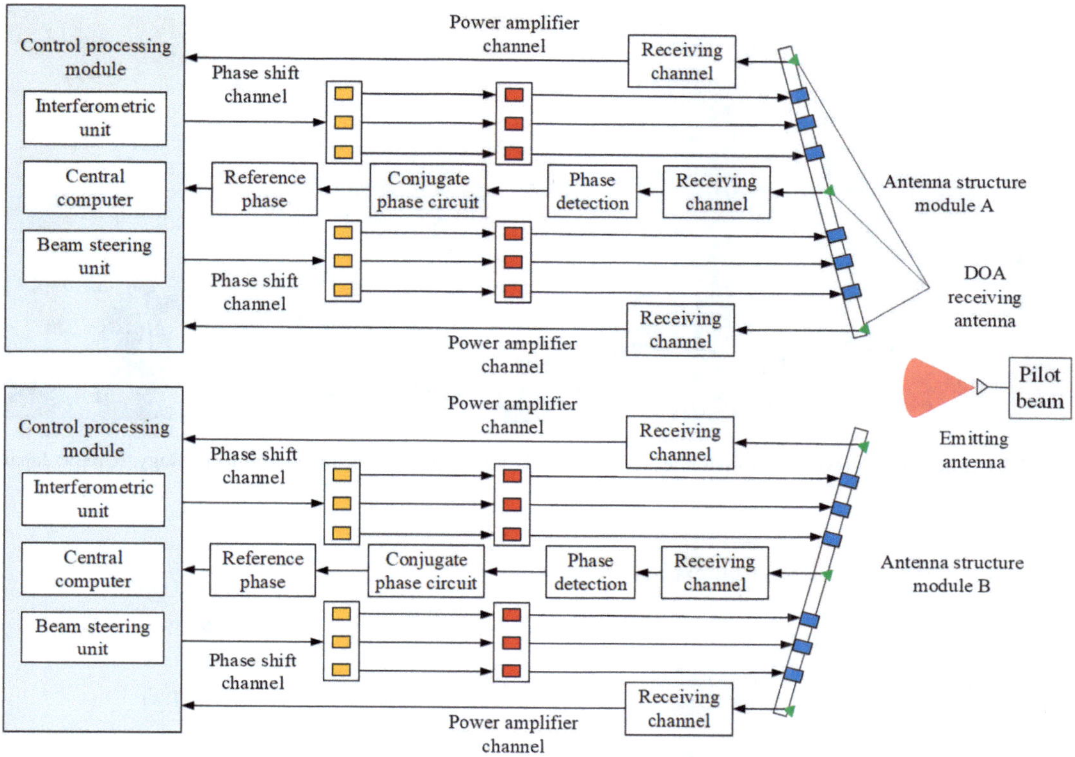

Fig. 5.20 Schematic diagram of high-precision microwave beam direction control based on pilot beam deviation compensation

the microwave receiving end, the pilot beam receiving antenna in the center of each antenna structure module calculates the conjugate phase corresponding to the received signal through the pilot beam, and this phase is used as the reference phase of this antenna structure module. At the same time, multiple receiving antennas are arranged around each antenna structure module, and the attitude deviation of the antenna subarray relative to the reference position is calculated using the received signal through the interferometer angle measurement method. Based on the reference phase of the antenna structure module, the phase of each antenna unit inside the antenna structure module is adjusted through the beam steering unit to compensate for the position and attitude deviation of the antenna structure module, achieving high-precision and efficient power transmission. Figure 5.21 shows a microwave power transmission test device based on retrodirective beam control technology conducted by Japan in 2015.

5.2 Laser Wireless Power Transmission

5.2.1 Composition of Laser Wireless Power Transmission System

The laser wireless power transmission system mainly consists of a laser source, laser transmission and beam direction control device, laser receiving and conversion device, etc. The whole operation status of laser wireless power transmission system needs to be controlled and monitored, as shown in Fig. 5.22.

The laser source is used to generate laser of the special wavelength, required power and quality, mainly composed of a laser, a laser combiner, a laser drive power supply, and a cooling device. Among them, the main function of the laser is to convert electric energy to laser. The main function of the laser combiner is to combine the laser produced by multiple lasers into a single beam

Fig. 5.21 Microwave power transmission test device based on retrodirective beam control technology. (Credit: Japan Space Systems)

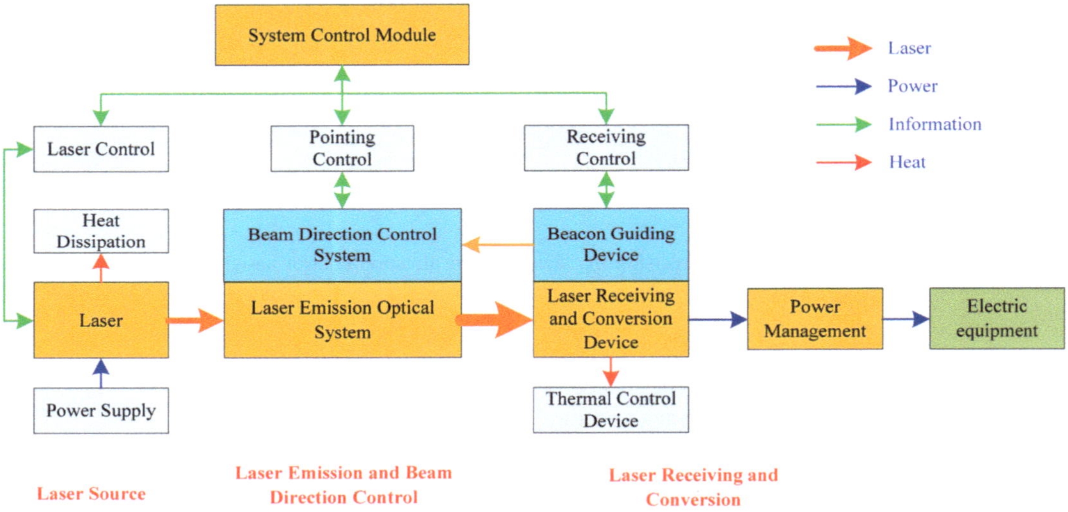

Fig. 5.22 Composition of laser wireless power transmission system

output. The laser drive power supply provides a constant current electric power for the laser, and the cooling device is responsible for maintaining a reasonable operating temperature of the laser.

The laser emission and beam direction control part mainly includes a laser emission optical system and a beam direction control system. Among them, the laser emission optical system is used to expand and collimate the laser output from the laser source, by adjusting the beam divergence angle and beam distribution to meet the needs of spot size, light power intensity, and spot uniformity at different distances. The beam direction control system is based on the capture and alignment of the receiving end, and the high-precision laser transmission is realized through the beam direction adjustment device.

The laser receiving and conversion part mainly includes a photoelectric receiving and conversion device, a power management unit, and a direction guidance device. Among them, the photoelectric reception and conversion device generally uses a laser PV array or a thermoelectric conversion device can be used. The power management unit mainly handles the unevenness and instability of the incident laser spot and the power output characteristics of the laser cell to achieve reliable and stable power supply to the load by performing

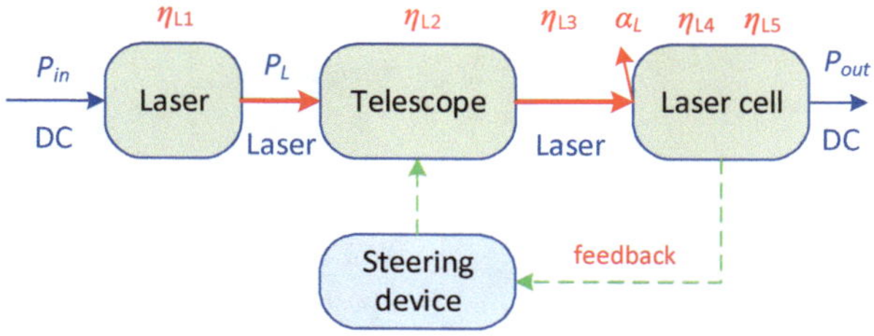

Fig. 5.23 Efficiency chain of laser wireless power transmission system. P_{in}—Input electric power, P_L—Output laser power, P_E—Laser transmission power, P_{out}—Output electric power, η_{L1}—Laser efficiency, η_{L2}—Laser emitting efficiency, η_{L3}—Laser transmission efficiency, η_{BCE}—Laser interception efficiency, η_{L4}—Laser PV array filling rate, η_{L5}—Laser cell conversion efficiency, η_{L6}—Electrical regulation efficiency

Fig. 5.24 Semiconductor laser chip

maximum power tracking on the laser PV array and configuring the corresponding conversion circuit according to the power supply demand of the load. The direction guidance device is used to emit a beacon beam to the beam direction control module to achieve high-precision laser beam transmission.

5.2.2 Efficiency Chain of Laser Wireless Power Transmission System

The efficiency chain of the laser wireless power transmission system is shown in Fig. 5.23.

Its main influencing factors include laser efficiency, laser transmission efficiency, laser interception efficiency, and laser cell conversion efficiency. Among them, laser efficiency and laser cell conversion efficiency are related to the device, laser transmission efficiency is related to atmospheric quality, and laser interception efficiency is related to laser divergence angle, transmission distance, pointing accuracy, and the size of the receiving PV array.

5.2.3 High-Power Laser

The main factors to consider when choosing a laser include the output power of the laser, the electro-optical conversion efficiency, the central wavelength, the loss of the transmission medium, etc. The commonly used laser wavelengths are mainly 808 and 1064 nm. The lasers mainly include semiconductor lasers, fiber lasers, direct solar-pumping lasers, etc.

1. Semiconductor Laser

   The semiconductor laser is a laser with semiconductor material (Fig. 5.24) as the working medium. Its working principle is that through the pump method of electrical injection, between the conduction and valence bands of the semiconductor material, or between the donor and acceptor energy levels of the semiconductor material, the non-equilibrium carriers realize particle number inversion. Because the electron-hole pairs in the high-energy state are not stable, the particles in the high-energy state will return to the low-energy state, resulting in a large number of electron-

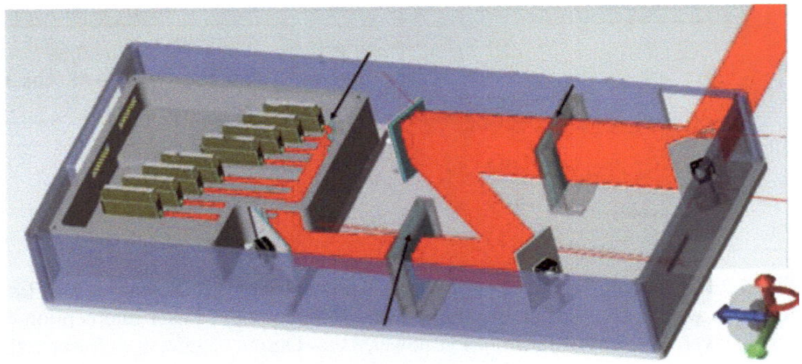

Fig. 5.25 Schematic diagram of spatial beam combination of multiple laser diodes. (Credit: Evergreen)

hole recombination, realizing the emission of the laser. Since semiconductor lasers usually use direct bandgap materials as working mediums, the energy level transition process does not involve phonon energy exchange, so the electro-optical conversion efficiency of semiconductor lasers is very high, and the current highest conversion efficiency has exceeded 70%. Because the output power of a single semiconductor laser chip is limited, in order to obtain high-power laser output, methods such as spatial beam combining of multiple laser diodes or fiber beam combining (Fig. 5.25) can achieve high-power laser output of several kilowatts or more.

The laser wavelength of the semiconductor laser is related to the material. Generally, III-V compound materials are used, and two types of material systems are commonly used. One is based on GaAs and $Al_xGa_{1-x}As$ (where x represents the percentage of Ga atoms in GaAs replaced by Al atoms), the value of x directly affects the laser wavelength, generally around 850 nm. The other is based on InP and InGaAsP, with a laser wavelength generally between 920 and 1650 nm. In order to match the receiving end laser cell, the wavelength of the semiconductor laser used for wireless power transmission is generally 808 nm.

The semiconductor lasers typically use an optical resonant cavity as a medium waveguide cavity, and the size of the resonant cavity mirror of the semiconductor laser is small, so the directionality of the output laser is poor. Moreover, due to the significant difference in the thickness and stripe size of the active area of the semiconductor laser, the horizontal and vertical divergence angles of the emitted laser beam are significantly different. The vertical divergence angle is usually much larger than the horizontal divergence angle. Because high-power semiconductor lasers use multiple semiconductor laser diodes for laser beam combining, the performance differences of different semiconductor laser diodes also make the beam directionality of the output of high-power semiconductor lasers, the divergence angle, and the beam energy distribution uniformity poor. Therefore, the quality of the output laser beam of the semiconductor laser is poor and is not suitable for long distance laser power transmission.

2. Fiber Laser

The fiber laser is the same as the normal laser, which also includes a pump source, a gain medium, and a resonant cavity. Among them, the pump source provides energy, stimulates the working material to achieve particle number inversion distribution; the gain medium is a fiber made of different materials, which produces laser by absorbing pump light to stimulate radiation; the resonant cavity enhances the stimulated radiation intensity through optical positive feedback and realizes the laser output of a specific wavelength.

The pumping method usually used by fiber lasers is optical pumping. The optical pump source, as the core device of the fiber laser, plays a decisive role in the performance of the laser. With the rapid development of semiconductor laser technology, it has become the preferred pump source for fiber lasers. In addition to considering factors such as power and efficiency in the selection of the pump source, it is more important to select a specific wavelength of the semiconductor laser in combination with the characteristics of the gain medium. For example, for a fiber gain medium doped with ytterbium ions, 915 and 975 nm are the two absorption peaks of ytterbium particles, so a 915 or 975 nm laser diode (975 nm corresponding spectral line is narrow, considering the temperature effect on the wavelength of the semiconductor laser, so the temperature requirement is high) is selected as the pump laser diode.

The gain medium is a fiber doped with different rare earth ions. When the pump light passes through the doped fiber, it will be absorbed by the rare earth ions. At this time, the electrons of the rare earth ions that have absorbed the photon energy will transition to a higher energy level, thereby achieving particle number inversion. The inverted particles will then transition from the high-energy level to the ground state in the form of radiation. And it releases energy, achieving stimulated radiation, producing radiation of the same frequency and phase, thus forming a laser with good coherence. The core of the general gain fiber is doped with rare earth elements such as ytterbium (Yb), erbium (Er), neodymium (Nd), and thulium (Tm). Because different rare earth ions have different energy level structures, when pump light is shot into different types of rare earth ions, it will excite lasers of different wavelengths. The fiber laser wavelength of doped ytterbium ions is 975 m or 1010–1200 nm, which is the wavelength chosen for laser wireless power transmission, with a good atmospheric transmittance. In order to obtain a high beam quality laser output, the core of the laser transmission is very thin, with a diameter generally ranging from a few micrometers to tens of micrometers, so it is very difficult to directly couple the pump light into the core. To solve this problem, a double-clad fiber structure is generally used, as shown in Fig. 5.26. The center of the fiber is the core, forming the waveguide structure for laser transmission. The second layer is the inner cladding, with a lower refractive index, and its diameter can reach hundreds of micrometers. The third layer is the outer cladding, its refractive index is lower than the inner cladding, which can cause total internal reflection when the pump light is transmitted to the interface of the inner and outer cladding, realizing the transmission of the pump light within the range of the inner cladding. The outermost layer of the fiber is the coating layer, which mainly plays a protective role. Therefore, the pump light does not need to be directly coupled into the smaller diameter core, but only needs to be first coupled into the larger diameter inner cladding, then transmitted along the fiber. When it reaches the interface of the inner and outer cladding, it will be totally reflected, and will be transmitted again within the range of the inner cladding. For the inner cladding of special shape, the pump light will pass through the core multiple times and be absorbed by the rare earth ions in the core, thus producing laser output.

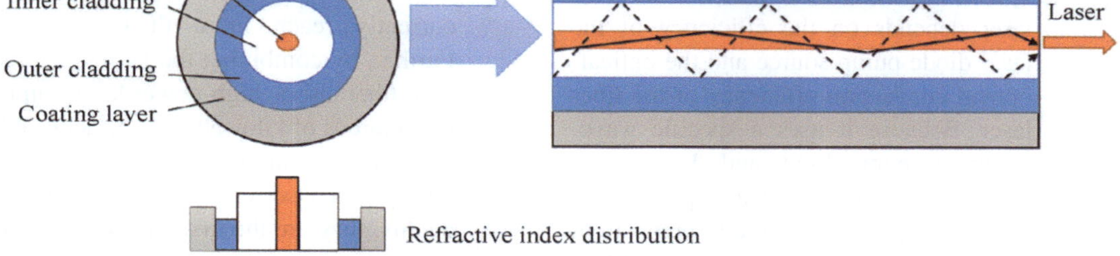

Fig. 5.26 Schematic diagram of double cladding fiber structure

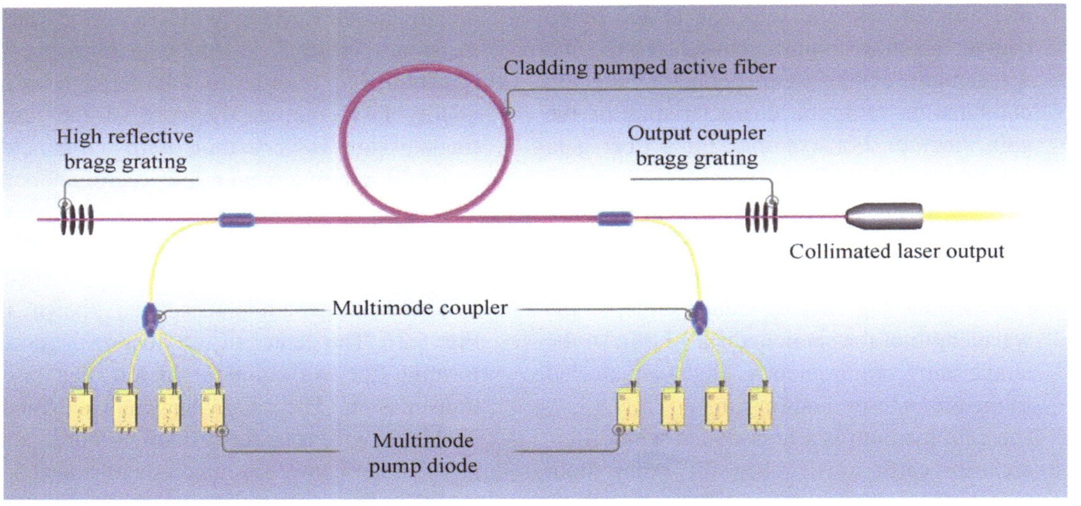

Fig. 5.27 Schematic diagram of bragg grating fiber laser based on laser diode pumping

The resonant cavity of the fiber laser includes various forms, typical ones are: Fabry-Perot resonant cavity (F-P), that is, using two cavity mirrors with different reflectivity for the pump light and laser to selectively transmit and reflect the pump light and laser, realizing laser output; Bragg grating, using specially designed grating to selectively transmit and reflect the pump light and laser, realizing laser output (Fig. 5.27). The main advantages of fiber lasers over semiconductor lasers are as follows:

(a) High beam quality. The beam quality of the laser depends on the number of laser modes transmitted in the core, which is directly related to the core diameter. Single-mode fiber lasers can achieve very high beam quality, and high-power laser output with high beam quality can also be obtained by combining multiple single-mode fiber lasers.
(b) High conversion efficiency. The electro-optical conversion efficiency of the fiber laser depends on the efficiency of the laser diode pump source and the optical-optical conversion efficiency of the fiber laser. Because it uses a specific wavelength of pump light, and through the double-clad structure of the special gain fiber, high-power multimode pump light is efficiently coupled into the gain fiber, making the optical-optical efficiency of the fiber laser higher, and the overall electro-optical conversion efficiency can reach 40–50%.
(c) Compact structure. Using a fiber coupler, the laser diode pump source can be fused with the gain fiber. At the same time, using a fiber grating to form a resonant cavity can make the structure of the laser fully fiberized, compact, stable in performance, and strong in environmental adaptability.
(d) Good heat dissipation. Due to the large surface area of the fiber, the heat generated by the fiber laser during operation can be dissipated more easily.
(e) Easy to achieve high power. Due to the unique structure of the fiber laser, the high-power pump source can be efficiently coupled into the gain fiber (Fig. 5.28), and at the same time, due to the good heat dissipation performance of the fiber laser, high-power single-mode laser output can be achieved, which can currently reach more than kilowatts. Further, by combining the lasers of multiple fiber lasers, high-power laser output of hundreds of kilowatts can be achieved. Because fiber lasers are easy to achieve high-power laser output with high beam quality, they are the first choice for high-power lasers in long distance transmission applications.

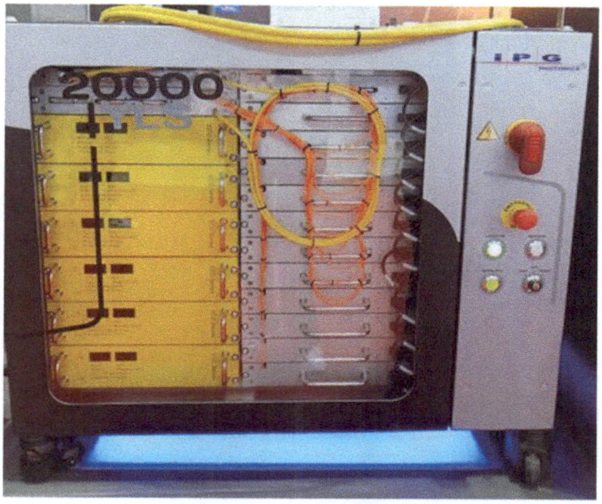

Fig. 5.28 Multiple laser diode modules coupled pump fiber laser

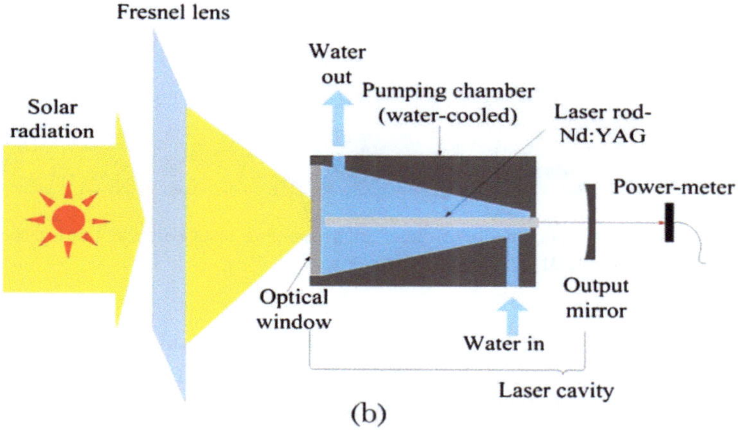

Fig. 5.29 Principle of direct solar-pumping laser

3. Direct solar-pumping lasers

A direct solar-pumping laser refers to a device that uses sunlight as a pump source, concentrates sunlight to directly pump the laser medium, and converts sunlight into a laser. Its biggest technical feature is to realize the direct conversion from sunlight to laser, thereby simplifying the conversion process from light to electricity, and then from electricity to light. As shown in Fig. 5.29, the incident sunlight is converged by a primary concentrator to a focusing cavity installed at the focal point. The focusing cavity realizes secondary focusing through reflection which maximizes the coupling of sunlight into the laser medium and improves the pumping efficiency. At the same time, it enables the laser medium to obtain uniform pumping, generates stimulated radiation and forms stable resonance, and achieves laser output through the laser output mirror.

In order to simplify the structure of the primary concentrator, the Fresnel lens method is generally used. The laser medium generally uses Nd:YAG crystals and Cr:Nd:YAG ceramics, which are made into rods to match the focusing cavity, producing 1064 nm laser. In order to ensure the normal operation, both the focusing cavity and the laser medium need to be well cooled, and a liquid cooling device with highly cooling capacity is required. The direct solar-pumping laser requires high precision for the direction of sunlight incidence, so a high-precision sun tracking system is needed. Currently, direct solar-pumping lasers (Fig. 5.30) are mainly in the research stage

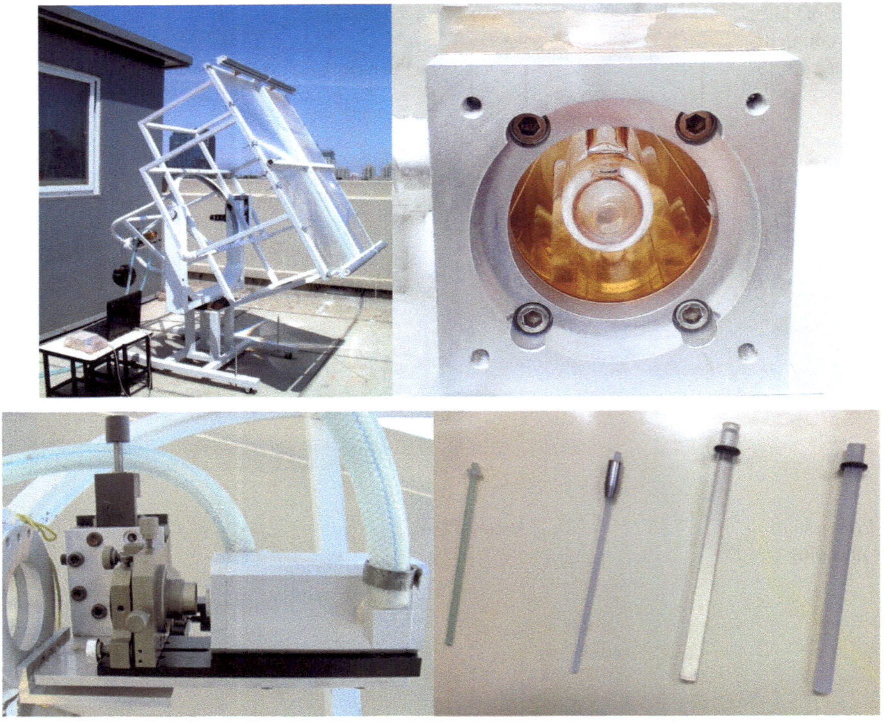

Fig. 5.30 Main components of direct solar-pumping laser (Fresnel Lens, Focusing Cavity, Cooling Pipeline, Laser Medium). (Credit: BIT)

and have not been practically applied. The main problem is that although the intermediate conversion of light- electricity-light is omitted, the efficiency of the direct solar-pumping laser is low with the highest level less than 10% currently. The laser power is low and the beam quality is poor. The complex cooling system is also the main factor limiting the application of direct solar-pumping lasers.

5.2.4 Laser Emission System

The laser output is generally a Gaussian beam with a certain divergence angle. In order to achieve long distance laser power transmission, the laser emission system needs to collimate the laser output from the laser source through an optical device to compress the divergence angle as much as possible, and to reduce the spot size at the receiving end. To meet the requirements of spot size and laser power intensity at different energy receiving distances, the laser emission system should also have a certain divergence angle adjustment capability. So, a focusing device needs to be designed to adjust the beam divergence angle. For applications with special requirements for high spot uniformity, special optical devices need to be designed to achieve a more uniform light density of non-Gaussian distribution. At the same time, for long distance laser power transmission, the pointing accuracy of the beam is particularly important, especially for the moving receiver. The laser emission system needs to capture and align with the receiver and achieve high-precision laser emission through the beam pointing adjustment device. Figure 5.31 shows the laser emission optical system used by the Japan Aerospace Exploration Agency (JAXA) for laser wireless power transmission.

The laser output beam has a minimum diameter and the corresponding position is called the beam waist. The diameter becomes larger and larger after deviating from the beam waist, and the divergence angle far from the beam waist is approximately constant. Although the beam waist and far-field divergence angle can be changed by the optical system, for a certain output laser

Fig. 5.31 Laser emission optical system. (Credit: JAXA)

beam, the product of the beam waist radius and the far-field half divergence angle (Beam Parameter Product, BPP) is a constant, that is, the smaller the beam waist diameter, the larger the divergence angle, and the larger the beam waist diameter, the smaller the divergence angle. Therefore, to compress the divergence angle of the output laser beam to achieve collimation, it is necessary to expand the laser beam and increase the beam waist diameter of the laser beam. The beam expander system transforms the input narrow beam laser into a wide beam laser output through different optical designs. Assuming the beam expansion ratio is m, the corresponding laser beam divergence angle will be compressed to the original 1/m. Therefore, after the beam expander system, the laser can transmit a longer distance without exceeding the maximum beam diameter limit (Fig. 5.32).

Laser beam expander systems are mainly divided into transmission and reflection types. The transmission type beam expander system can be divided into Galilean beam expander system and Kepler beam expander system (Figs. 5.33 and 5.34), and the reflection type beam expander system generally includes Gregory system and Cassegrain system. Similarly, by adjusting the focal length of the optical system, the divergence angle of the output laser can be adjusted. For a more uniform receiver beam distribution, it is necessary to use multiple laser beam inputs (or laser beam splitting) corresponding to multiple optical systems for beam combination.

For a 1064 nm laser, assuming that the beam quality of a single fiber laser is about 1.2, and the beam quality of a high-power laser formed by spectral beam combining technology is about 1.8, corresponding to different diameters of the emission optical system and 10 times focusing ability, the size of the receiver beam spot at different transmission distances is shown in Table 5.1. Among them, the spot diameter corresponds to the position of 86.5% energy interception, the distance of 40,000 km corresponds to the minimum divergence angle, and all spot sizes do not exceed the spot size corresponding to 40,000 km.

Laser wireless power transmission requires high-precision beam pointing control, which can adopt APT similar to laser communication, namely Acquisition, Pointing, and Tracking (APT). The basic principle is that the receiving end emits a beacon laser, then the emitting end captures the beacon laser, and through high-precision mechanical and optical pointing control, continuous tracking of the receiving end is achieved to realize continuous high-precision power transmission. APT systems are generally divided into coarse tracking subsystems, fine tracking subsystems, and signal processing and control subsystems. The coarse tracking subsystem is a dual-axis gimbal that can drive the optical system for a wide range of angle adjustments, but the positioning accuracy is low, mainly for large field of view capture and tracking. The fine tracking subsystem is based on the coarse tracking subsystem, according to the captured beacon laser signal, by using a high-precision swing mirror integrated in the optical system driven by piezoelectric ceramics or voice coil motors to

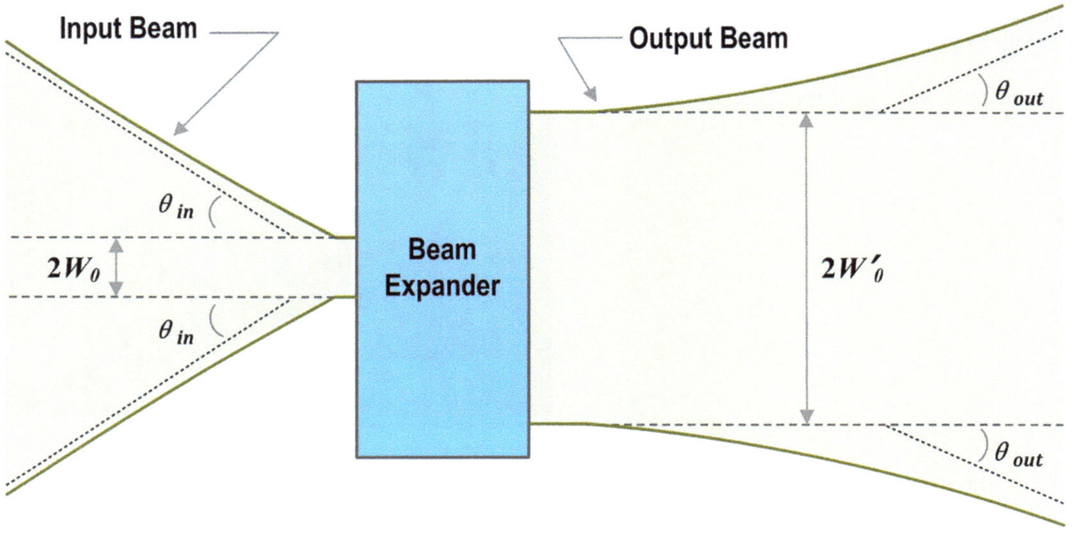

Fig. 5.32 Schematic diagram of laser beam expansion

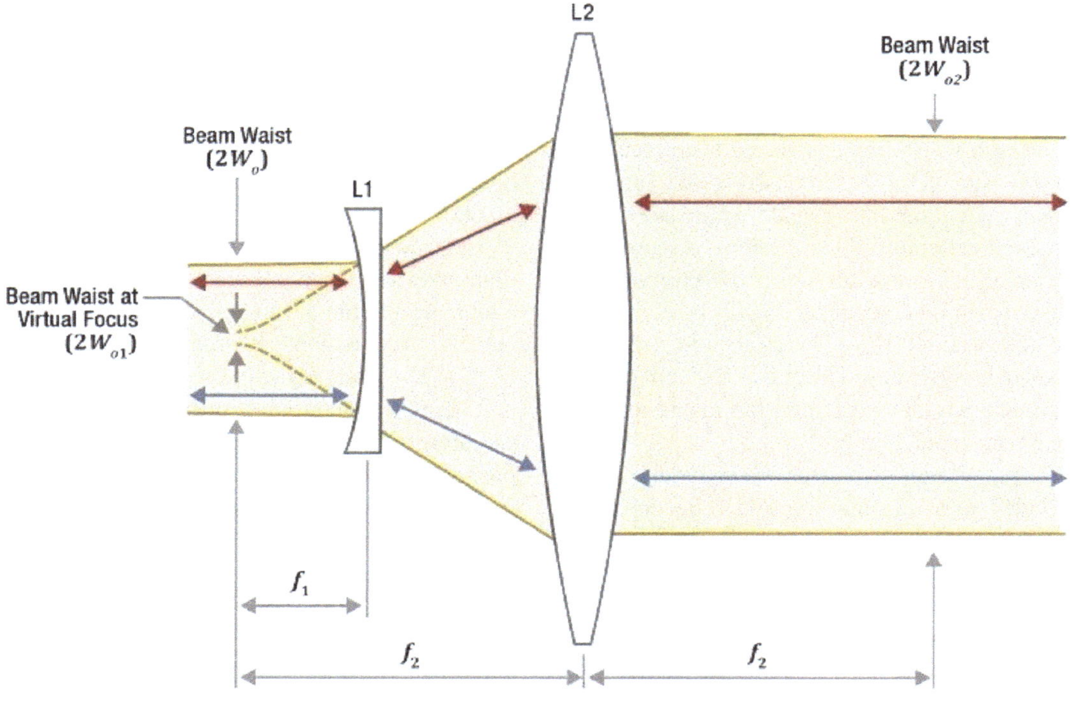

Fig. 5.33 Galileo beam expansion system

achieve precise pointing and fine tracking of the target. The signal processing and control subsystem accurately controls the coarse and fine tracking subsystems based on attitude information and beacon laser information fed back by CCD sensors.

5.2.5 High Efficiency Laser Reception and Conversion

The receiving end of laser wireless power transmission should collect the laser beam as much as possible and efficiently convert it into electric

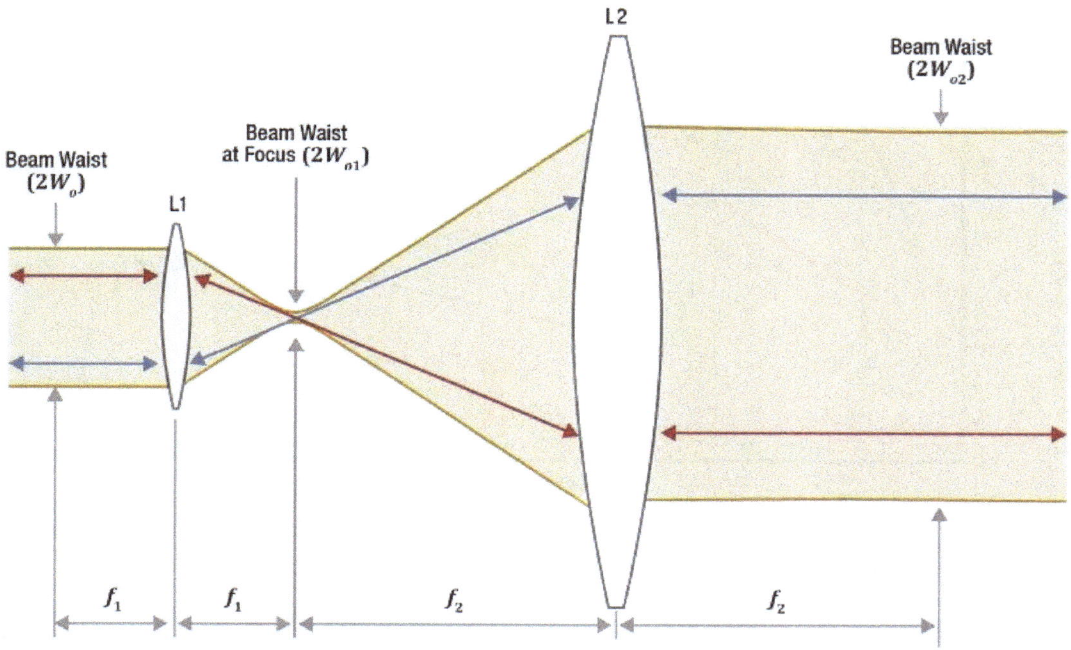

Fig. 5.34 Kepler beam expansion system

Table 5.1 Relationship between 1064 nm laser receiver spot diameter (m) and transmission distance for various optical diameters

	Transmission distance/km				
Optical diameter/mm	100	1000	10,000	20,000	40,000
1000	1.03–10.3	2.6–26.4	24.4–97.5	48.8–97.5	97.5
2000	2.0–20.0	2.3–23.4	12.4–48.8	24.5–48.8	48.8
3000	3.0–30.0	3.1–31.1	8.7–32.7	16.5–32.7	32.7
5000	5.0–20.1	5.0–20.1	7.0–20.1	11.0–20.1	20.1

energy to power the load. To efficiently collect the laser, the laser receiving end should cover the entire laser spot (since the laser beam is Gaussian distributed, generally considering the position corresponding to 86.5% energy collection). Considering the position deviation caused by laser pointing accuracy, the area of laser receiving end should be larger than the area of laser spot. For ultra-long distance power transmission, due to the limitations of the size of the optical emitting end and receiving end, it should cover as much as possible the spot part where the beam center energy is higher. At the same time, the laser cell should have a very high laser absorption rate. The high efficiency conversion of the laser can also consider the light-heat-electricity method, but generally the light-electricity method is used. Compared solar cells, the laser spectrum range of laser cells is extremely narrow, while the spectrum range of sunlight is wide. In addition, the distribution of lasers is extremely uneven Gaussian distribution, while sunlight is uniformly distributed.

Because the laser is monochromatic or quasi-monochromatic radiation, by choosing semiconductor materials with appropriate bandgap widths (bandgap widths slightly smaller than photon energy and high quantum efficiency), the photon energy can be maximized, theoretically achieving a photoelectric conversion efficiency far higher than that of solar cells. Figure 5.35 shows the photoelectric conversion efficiency of some

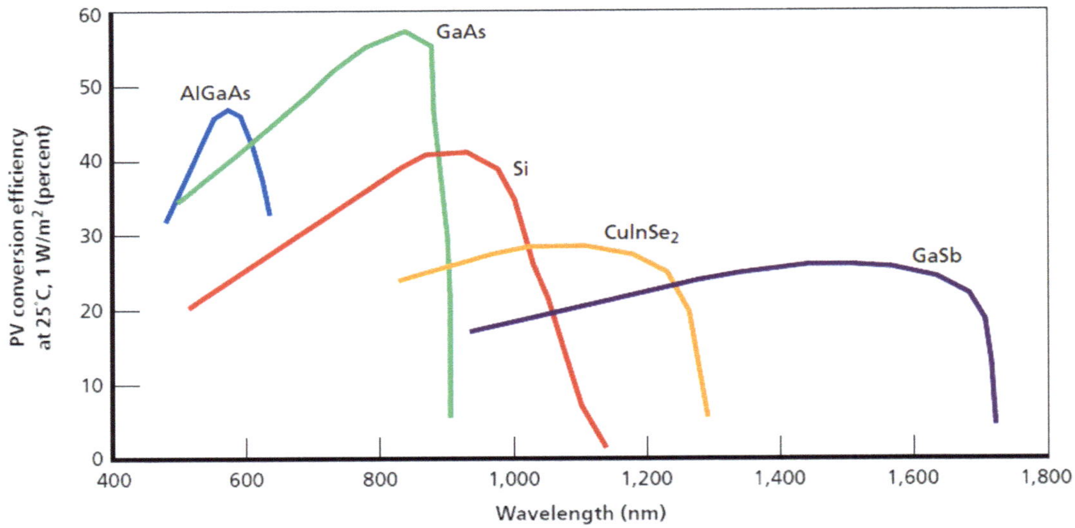

Fig. 5.35 Photoelectric conversion efficiency of different semiconductor materials at different wavelengths

semiconductor materials at different wavelengths. It can be seen that the photoelectric conversion efficiency of silicon-based laser cells is relatively low, but a conversion efficiency of 30% can be achieved for wavelengths of 700–1000 nm. GaAs-based laser cells have a high photoelectric conversion efficiency in the wavelength range of 800–850 nm, which matches the semiconductor laser around 800 nm and is currently the focus of research for near and medium distance transmission. The Fraunhofer ISE in Germany has developed a GaAs-based laser cell with an area of 0.054 cm^2, using an 858 nm laser, and achieved a photoelectric conversion efficiency of 68.9% at an average radiation intensity of 11.4 W/cm^2, which is the highest experimental efficiency reported so far (Fig. 5.36). However, when the wavelength exceeds 870 nm, the efficiency of GaAs-based laser cells drops rapidly. For the output laser above 1000 nm common in fiber lasers, III-V semiconductor materials $In_xGa_{1-x}As$ are used to reduce the bandgap width of the material (as the In content increases, the bandgap width decreases) and to achieve spectral response to longer wavelength lasers, as shown in Fig. 5.37. Currently, under the laser irradiation with a wavelength of 1064 nm and an average radiation intensity of 5.2 W/cm^2, the efficiency of the $In_{0.23}Ga_{0.77}As$-based laser cell reaches 50.6%. The design of laser cells should also consider practical applications. For high-power intensity lasers, the device design under high-power intensity and special heat dissipation design should be particularly considered. For extremely low-power intensity laser applications, the spectral response of laser cells under weak light should be considered.

In addition to selecting matching laser cells because the laser beam is a highly uneven Gaussian distribution (Fig. 5.38), considering the change in laser pointing, the laser power intensity on the surface of the laser PV array is extremely uneven and dynamically changes greatly which make the electrical connection of the laser cells very complex and greatly affects the conversion efficiency of the receiving PV array. Under uneven radiation, direct series connection of cells will cause serious mismatch between cells, thereby reducing the conversion efficiency of the array. And the direct parallel connection of cells will make the output voltage of the array low. While the light intensity distribution changes, the stability of the array efficiency becomes poor. Therefore, the core idea of improving the conversion efficiency of the photovoltaic array by optimizing the electrical connection structure of the photovoltaic array is: under the premise of meeting the system

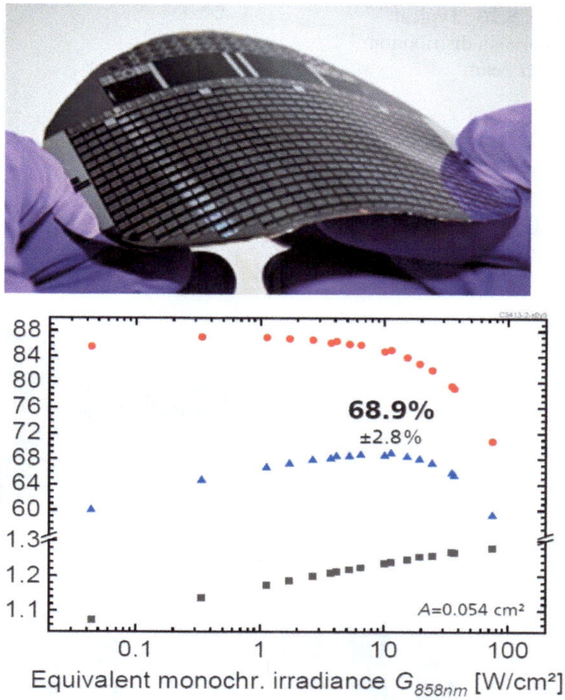

Fig. 5.36 High efficiency laser cell developed by Fraunhofer ISE Institute. (Credit: Fraunhofer ISE)

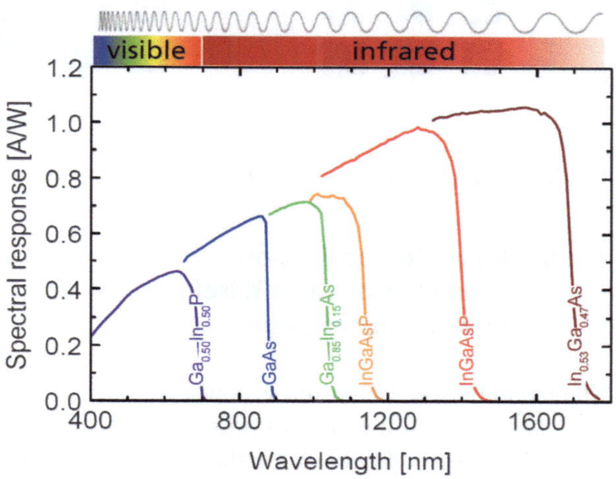

Fig. 5.37 Spectral response curves of different semiconductor materials corresponding to different wavelengths. (Credit: Fraunhofer ISE)

working voltage and current, adjust the distribution of uneven illumination in the electrical structure of the photovoltaic array through different series and parallel methods, including series-parallel (SP) structure, total cross-tie (TCT) structure, and bridge-linked (BL) structure (Fig. 5.39). Among them, the TCT structure is a kind of electrical connection method of laser cells that is parallel first and then series, which can better solve the power loss problem caused by cell mismatch when the radiation is uneven, thereby minimizing the mismatch phenomenon and improving the conversion efficiency of the array. For practical applications, how to make reasonable electrical connections and achieve high efficiency power output with a high dynamic range through control still faces a great challenge.

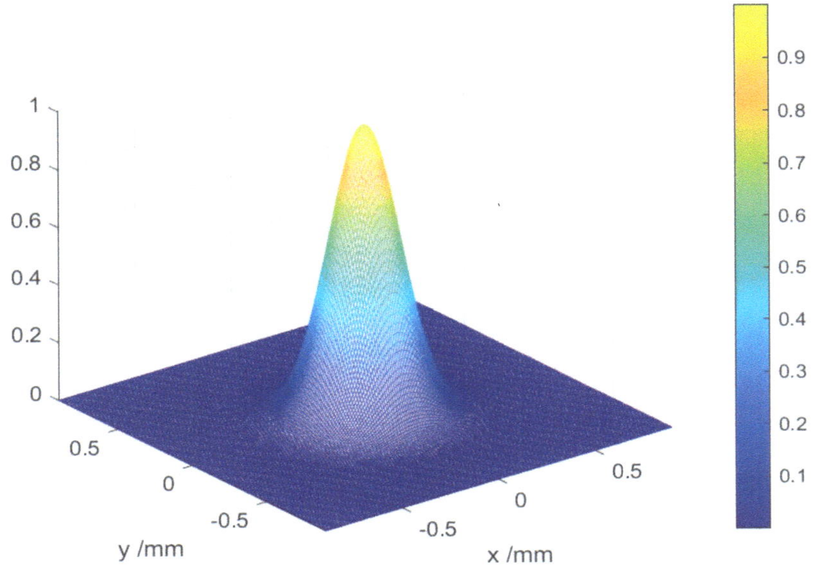

Fig. 5.38 Typical Gaussian distribution laser beam

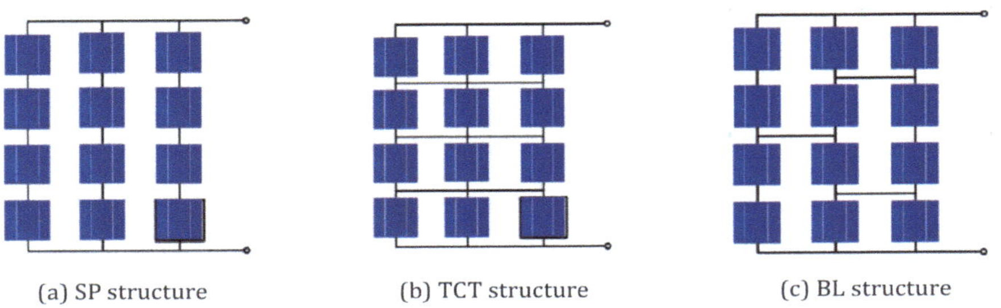

Fig. 5.39 Typical laser cells connection method. (**a**) SP structure, (**b**) TCT structure, (**c**) BL structure

5.2.6 Possible Application Scenarios of Laser Wireless Power Transmission

The flexibility of laser wireless power transmission makes it have a variety of possible applications (Fig. 5.40). Figure 5.41 shows different application scenarios of laser wireless power transmission. The emission end location can be ground, air, stratosphere, space (low orbit, high orbit), lunar orbit, lunar surface, etc. and the receiving end location mainly includes ground, air, space, lunar surface, etc. Table 5.2 lists the technical requirements and characteristics corresponding to different application scenarios.

5.2.7 Lunar Laser Wireless Power Transmission

Lunar exploration and lunar resource exploitation and utilization are the key development directions in the future aerospace field. Energy supply has become a key issue restricting lunar exploration and lunar resource exploitation and utilization, especially the power supply during the 14-day lunar night is a major challenge. Using a high-power laser wireless power transmission system on the lunar orbit or Earth-Moon L1/L2 orbit to supply energy to the lander, lunar rover, and lunar base will greatly increase the working time and activity range of the lunar probe on the

Fig. 5.40 Typical application of laser wireless power transmission. (Credit: ESA/NASA/JAXA)

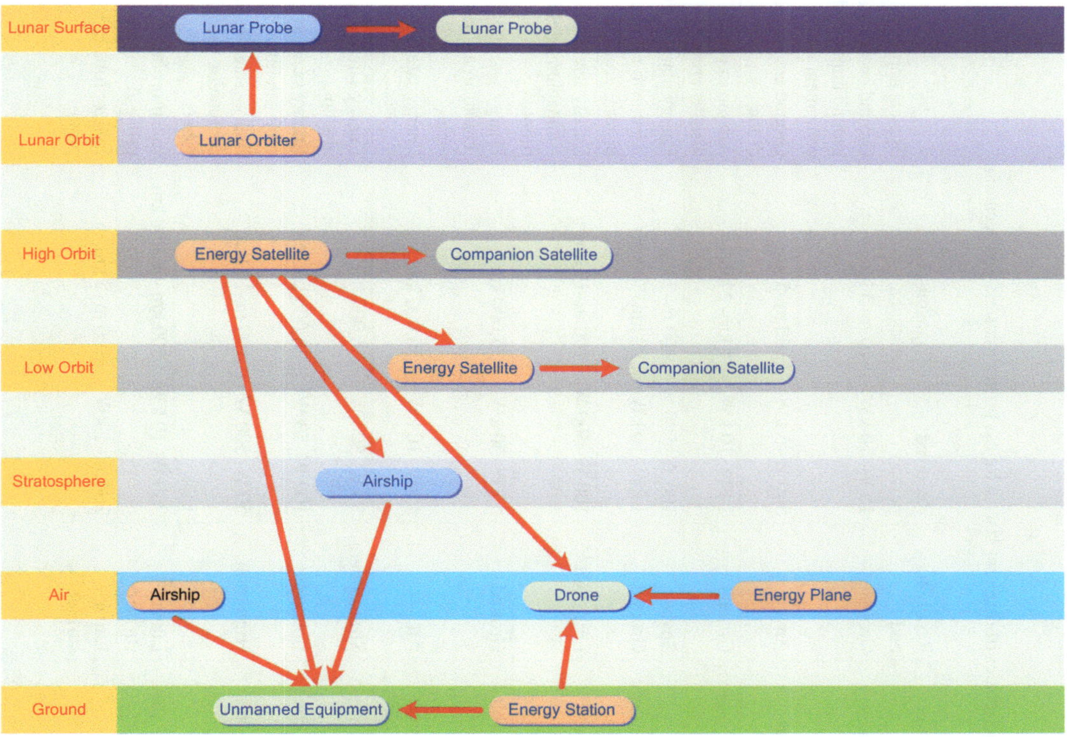

Fig. 5.41 Various possible scenarios of laser wireless power transmission

Table 5.2 Characteristics of different laser wireless power transmission scenarios

Serial number	Emission end	Receiving end	Received power	Transmission distance	Typical applications	Technical features			
						Power	Transmission distance	Directional range	Continuity
1	Ground	Ground	10 W–10 kW	10 m–5 km	Ground energy station to ground unmanned equipment	Medium-small	Medium-short	Wide	Weather dependence
2	Ground	Air	500 W–5 kW	100 m–5 km	Ground energy station to drone	Medium-small	Medium-short	Wide	Weather dependence
3	Air	Ground	1 kW–10 kW	1 km–2 km	Airship to remote areas, disaster areas	Medium-small	Medium	Narrow	Continuous
4	Air	Space	100 W–500 W	100 m–2 km	Energy plane to drone	Small	Medium-short	Narrow	Continuous
5	Stratosphere	Ground	1 kW–10 kW	20 km–30 km	Airship to remote areas, disaster areas	Medium-small	Long	Narrow	Weather dependence
6	Low orbit	Low orbit	100 W–500 W	100 m–10 km	Energy satellite to companion satellite, robots	Small	Medium-short	Narrow	Continuous
7	High orbit	Ground	5 kW–50 kW	36,000 km	Energy satellite to remote areas, disaster areas	Medium-large	Ultra-long	Narrow	Weather dependence
8	High orbit	Air	1 kW–10 kW	36,000 km	Energy satellite to drones	Medium-small	Ultra-long	Narrow	Weather dependence
9	High orbit	Stratosphere	5 kW–50 kW	36,000 km	Energy satellite to airships	Medium-large	Ultra-long	Narrow	Continuous
10	High orbit	Low orbit	100 W–5 kW	36,000 km	Energy satellite to low orbit satellites	Medium-small	Ultra-long	Wide	Intermittent
11	High orbit	High orbit	50 W–500 W	100 m–10 km	Energy satellite to companion satellites, robots	Small	Medium-short	Narrow	Continuous
12	Lunar orbit	Lunar surface	200 W–10 kW	500 km–60,000 km	Lunar orbiter to lunar probes/lunar base	Medium-small	Ultra-long	Wide	Intermittent/continuous
13	Lunar surface	Lunar surface	200 W–1 kW	100 m–5 km	Lunar probe to lunar probe	Small	Medium-short	Wide	Continuous

lunar surface. It will enhance the total energy supply required for lunar exploration and lunar resource exploitation and utilization, as well as the safety and flexibility of power supply. On the other hand, when exploring the permanent shadow areas of the south and north poles, which are special significance for lunar scientific exploration and resource exploitation and utilization, the laser wireless power transmission can be used to power the rover exploring the permanent shadow area of the moon that makes the exploration of the permanent shadow area possible and effectively improves the ability of lunar exploration.

For different lunar mission requirements, the main lunar exploration orbits include the lunar equatorial orbit, lunar polar orbit, lunar frozen orbit, and Earth-Moon L1/L2 halo orbit.

1. Lunar equatorial orbit: This mainly refers to the lunar orbit running around the lunar equator. The characteristic of this orbit is that there is a lunar shadow period and the length of the shadow period is related to the orbit height. The coverage area of this orbit on the lunar surface is mainly a certain area of south and north latitude centered on the equator, and each orbit can achieve coverage, but it cannot cover high latitudes and the south and north polar regions (Fig. 5.42).
2. Lunar polar orbit: This mainly refers to the lunar orbit passing through the south and north polar regions of the moon. The characteristic of this orbit is that it can be designed to have no lunar shadow period. This orbit can cover all areas of the moon, but only the south and north poles and nearby areas can be covered once in each orbit, and other areas need multiple orbits to cover once.
3. Lunar frozen orbit: This is a large elliptical orbit and its advantage is that it almost does not need to orbit maintain. It can be designed to cover a specific area for a long time, but it cannot achieve continuous coverage. The typical lunar frozen orbit period is 12 h, and each circle is visible for about 10 h.
4. Earth-Moon L1/L2 halo orbit: There are also five libration points in the Earth-Moon gravity system, similar to the Sun-Earth libration points. Among them, the distance from the Earth-Moon L1 point to the center of the moon is 53,000–59,000 km, and the distance from the Earth-Moon L2 point to the center of the moon is 59,000–66,000 km. Since the moon's rotation and revolution speeds are the same, the L1 point and L2 point are stationary relative to the lunar surface, the L1 point can cover almost all

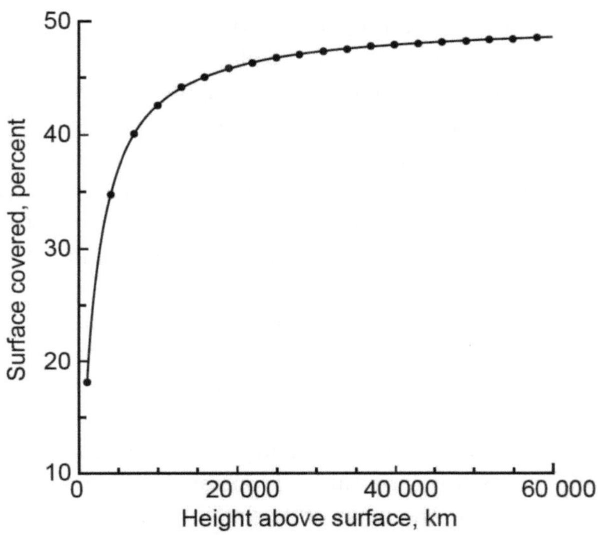

Fig. 5.42 Relationship between lunar orbit altitude and surface coverage

Fig. 5.43 Schematic diagram of lunar orbit laser wireless power transmission. (Credit: NASA)

of the front of the moon, and the L2 point can cover almost all of the back of the moon.

For a typical lunar probe power supply application based on laser wireless power transmission (Fig. 5.43), the preliminary concept is as follows.

The mission conceives an orbital laser wireless power transmission platform operating in the lunar equatorial orbit (or polar orbit) that provides discontinuous power supply to lunar surface explorers in the lunar equator (or polar region) and within a certain range of south and north latitudes during each orbital period to meet the survival and exploration needs. Assuming the transmitted power supply is 1 kW and the diameter of the receiving end is about 5 m. The mission includes the lunar orbiter and the lunar surface explorer, with the main specifications shown in Table 5.3. The orbiter operates in the lunar equatorial orbit, and the explorer operates within a latitude range of ±30°. The orbiter continuously tracks the sun, wirelessly transmits energy to the explorer through the laser. The explorer receives the laser and combines it with the battery to provide power for the payload, meeting the power supply needs for survival and operation during lunar nights. The lunar orbiter mainly includes the spacecraft platform, electric power generation system, high-power laser, and laser emission device. The power generated by the lunar orbiter is mainly used for battery power storage.

Table 5.3 Technical specifications of lunar Orbit laser power transmission

Lunar orbiter		Lunar surface explorer	
Laser transmission distance/km	500–900	Laser cell type	InGaAs
PV array power generation/kW	5	Laser cell conversion efficiency/%	40
Laser power supply/kW	12	Laser PV Array diameter/m	3
Laser wavelength/nm	1064	Power generation/kW	1.2
Laser electro-optical conversion efficiency/%	40	Average power supply/W	200
Laser emission power/kW	4.4		
Optical emission diameter/m	0.5		
Pointing accuracy/urad	1		

When the explorer is visible, the orbit will transmit laser power. The lunar surface explorer mainly obtains electric power from sunlight during the illumination period and obtains electric power from laser wireless power transmission during the lunar night period. Due to the discontinuity of laser transmission, the battery needs to be configured to provide basic energy supply for the explorer when it is in the shadow period and cannot receive laser energy.

Bibliography

1. Rodenbeck C T, Jaffe P I, Strassner II B H, et al. Microwave and Millimeter Wave Power Beaming[J]. IEEE Journal of Microwaves, 2021, 1 (1): 229-259.
2. Jin K, Zhou W Y. Wireless Laser Power Transmission: A Review of Recent Progress[J]. IEEE Transactions on Power Electronics, TPEL-Reg- 2018-02-0304.
3. Jaffe P, Hodkin J, Harrington F, et al. Sandwich Module Prototype Progress for Space solar power[J]. Acta Astronautica, 2014, 94: 662–671.
4. Hashemi M R, Fikes A C, Gal-Katziri M, et al. A Flexible Phased Array System with Low Areal Mass Density[J]. Nature Electronics, 2019, (5): 195-205.
5. Shi D, Hou X B, Huang X J, et al. Designing of Long Distance LWPT System for SPS[J]. Advances in Astronautics Science and Technology, 2022, 5 (1): 11-17.
6. Zhao C M. Solar Pumped Laser[M] (in Chinese). Beijing: National Defence Industry Press, 2016.
7. Huang K M, Chen X, Li C J. Principles and Technologies of Microwave Wireless Power Transmission[M] (in Chinese). Beijing: Science Press, 2021.
8. Ma H H, Li C G, Dong Y Z, et l. Space Wireless Power Transmission Technique[M] (in Chinese). Beijing: Beijing Institute of Technology Press, 2019.
9. Zhao C M, Wang Y S, Guo L D, et al. Development of laser wireless power transmission technology [J] (in Chinese). Laser Technology, 2020, 44(5):538-545.
10. Hou X B, Wang X, Wang L, et al. Space Power Satellite Retrodirective Beam Steering Simulation [J] (in Chinese). Journal of Astronautics, 2016, 37(7):887-894.
11. Wang K, Wang X C, Qian B, et al. Recent Development on Space Application for High-Efficiency Solar Cells and Array Technology[J] (in Chinese). Journal of the Chinese Ceramic Society, 2022, 50(5): 1436-1446.
12. Ni W, Liu X J. Research Progress on Laser Photovoltaic Cells[J] (in Chinese). Chinese Journal of Power Sources, 2022, 42(4):356-359.
13. Li J, Yu H, Yu T C et al. Design of High Efficiency Diode Laser Module for Wireless Power Transmission[J] (in Chinese). Infrared and Laser Engineering, 2021, 50(5):31-8.
14. Li Z Y, Zhang J D, Huang X J. Laser Wireless Power Transmission Technology for Solar Power Satellite[J] (in Chinese). Spacecraft Engineering, 2015, 24(1):31-37.
15. Hou X B, Shi D L, Xu H Y. Analysis on the Applications of Laser Power Transmission[C] (in Chinese). 5st Wireless Power Transmission and Energy Interconnection Forum, Chendu, China, December, 2021.
16. Li X. Study on the Problems of Large Transmitting Array Antennas for Space Solar Power Satellite[D](in Chinese). Xidian University, 2017.9.
17. Ruan B D. Retro-reflective Tracking and Focusing Technique for Microwave Power Transmission[D] (in Chinese). Nanjing University of Aeronautics and Astronautics, 2017.3.
18. Cao X M. Microwave Power Transmission Technique for Moving Targets[D](in Chinese). Nanjing University of Aeronautics and Astronautics, 2019.12.

6 Transportation, On-Orbit Construction, and End-of-Life Decommission of SPSs

6.1 Analysis of the Assembly and Transportation Mode of SPSs

The mass and volume of SPSs are enormous, and they need to be broken down into multiple modules for launch into orbit based on the transport capacity of the launch vehicle (including launch mass and envelope size). After on-orbit assembly, the final operating orbit is the GEO. Generally, launch vehicles are used to launch satellites into geosynchronous transfer orbit (GTO), and then the satellite's own propulsion subsystem completes the transfer from GTO to GEO. Alternatively, they can be directly launched into GEO or gradually transferred from LEO to GEO using a propulsion system (Fig. 6.1).

For SPSs, there are three typical mission modes based on the assembly location of the modules: the LEO assembly and transportation mode, the GEO assembly and transportation mode, and the combined assembly and transportation mode of LEO and GEO. These three modes have significant differences in their impact on the transportation process and capacity, but all require the use of heavy-lift launch vehicles to launch the SPS modules into LEO firstly. For the transfer from LEO to GEO, considering different assembly modes, the main differences in the corresponding transportation modes lie in the different payload masses transported each time and the capabilities of the inter-orbit transporters used.

6.1.1 LEO Assembly and Transportation Mode

First, use heavy-lift rockets to launch all the SPS modules into LEO, where the entire SPS is assembled. Then, rely on the propulsion system of the SPS itself to transport the SPS from LEO to GEO (Fig. 6.2). The main features include: all modules are assembled in LEO, the assembly difficulty is relatively low, and it is convenient for astronauts to participate in the assembly; the stay time in LEO is long, and the impact of atomic oxygen environment, space debris, etc. is significant; the aerodynamic resistance and gravity in low orbit are relatively large and the propellent consumed for attitude adjustment and orbit maintenance during assembly is large; the overall transportation requires the propulsion system with high thrust and amount of propellant, and the overall scale of the SPS will significantly increase; due to the limitation of the SPS's total thrust, the time for the SPS to move from LEO to GEO is long; because the SPS itself cannot increase radiation protection, the Earth's radiation belt has a significant radiation impact on the SPS's equipment, especially the performance degradation of the solar array is significant, which will greatly affect the life of the SPS; there is no need of the support of a dedicated inter-orbit transporter.

X. Hou, *Space Solar Power*, Sustainable Development Goals Series,
https://doi.org/10.1007/978-981-97-3580-8_6

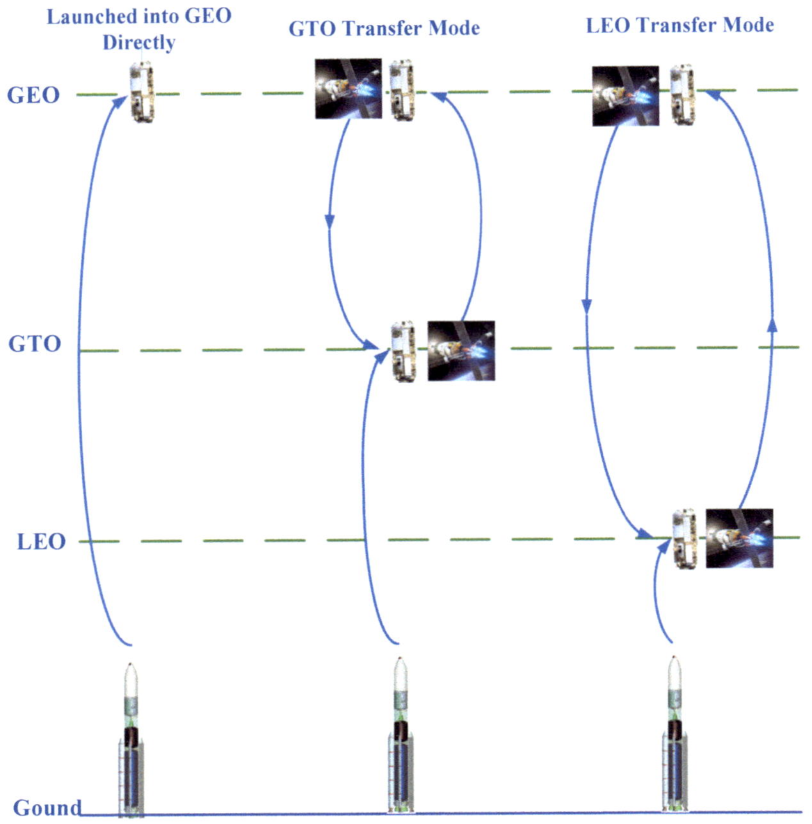

Fig. 6.1 GEO orbit satellite launch method

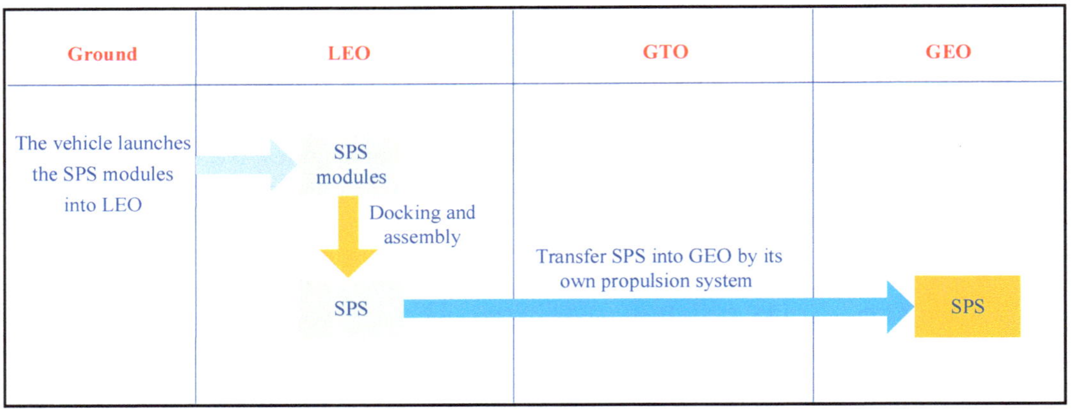

Fig. 6.2 Schematic diagram of LEO assembly and transportation mode

6.1.2 GEO Assembly and Transportation Mode

First, use heavy-lift rockets to launch the SPS modules into LEO, then use the upper stage of the rocket or an inter-orbit transporter to transport all the SPS modules to GEO, where all the modules are assembled (Fig. 6.3). The main features include: all assembly processes are carried out in GEO, so the impact of space debris is relatively small, and the possibility of astronauts participating in the assembly is small; modular

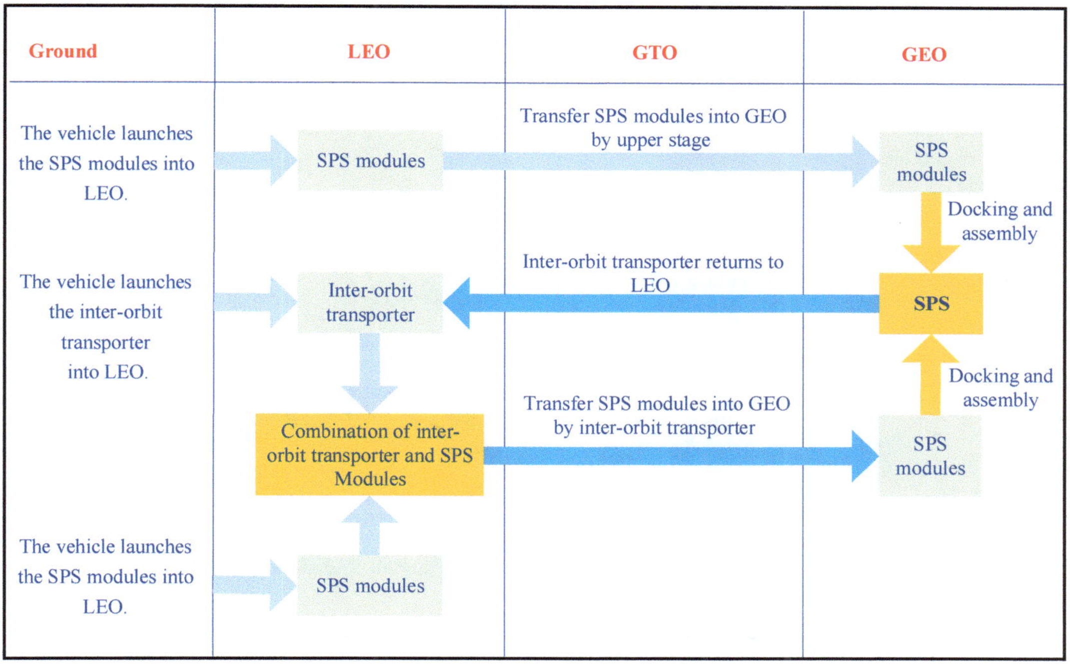

Fig. 6.3 Schematic diagram of GEO assembly and transportation mode

transportation is carried out, so the transportation capacity requirement of the inter-orbit transporter is relatively low, but more inter-orbit transporters are needed for support; the radiation impact of the Earth's radiation belt on the SPS modules during orbit transfer can be reduced by the protective shield of the inter-orbit transporter; the inter-orbit transporter can achieve long-term reusable through propellant refueling and replacement of the solar array in LEO.

6.1.3 Combination of LEO and GEO Assembly and Transportation Mode

Use heavy-lift rockets to send all the SPS modules directly into LEO and assemble some of the modules in low Earth orbit according to needs, then use an inter-orbit transporter to transport the partially assembled modules to GEO, and finally complete the final assembly of the SPS in GEO (Fig. 6.4). The main features include: the complex assembly is carried out in LEO, which is convenient for astronauts to participate in the assembly; the assembly time in LEO is short, and the impact of space atomic oxygen, space debris, etc. is relatively small; the transportation capacity requirement of the inter-orbit transporter is relatively high, but fewer inter-orbit transporters are needed; the SPS module assembly in LEO may be in a deployed state, it is difficult to protect it with a protective shield, and the radiation impact of the Earth's radiation belt on the SPS module assembly during orbit transfer is significant; the inter-orbit transporter can achieve long-term reusable through propellant refueling and replacement of the solar array in LEO.

6.2 Transportation of SPSs

Based on the analysis of the assembly and transportation modes above, the transportation of the SPS needs to consider the transportation from the ground to LEO and from LEO to GEO.

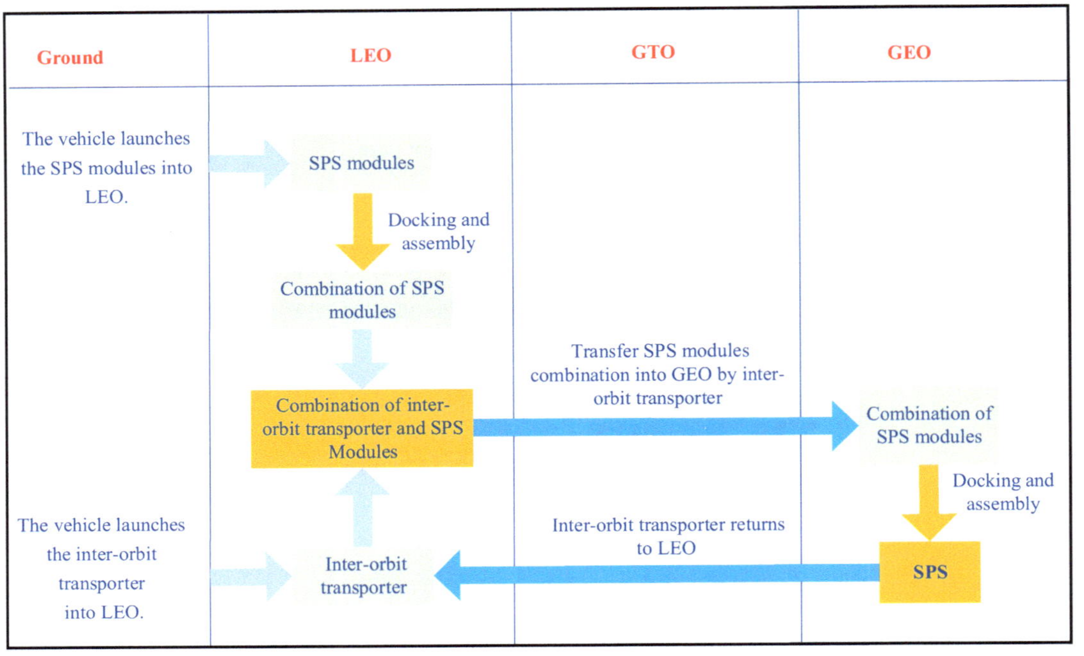

Fig. 6.4 Schematic diagram of combination of LEO and GEO assembly and transportation mode

6.2.1 Ground-LEO Transportation

Currently, the typical rockets in service or under development internationally that can be used for the launch of SPSs mainly include the Long March 5 launch vehicle, the Long March 9 heavy-lift launch vehicle, the Falcon Heavy Launch Vehicle, the Space Launch System, the Starship system, etc.

1. Long March 5 Launch Vehicle (CZ-5)

The Long March 5 series of launch vehicles is the most powerful rocket currently in service in China. It is the main launch vehicle for the construction of the space station and the launch of deep space explorers. It includes the basic model of the Long March 5 rocket (CZ-5) with a two-and-a-half stage configuration, the Long March 5B rocket (CZ-5B) with a one-and-a-half stage configuration, and the Long March 5/Yuanzheng 2 rocket (CZ-5/YZ-2) with an additional upper stage. Among them, the CZ-5 can be used for many launch missions such as geosynchronous transfer orbit, lunar transfer orbit, and Mars transfer orbit with a geosynchronous transfer orbit carrying capacity reaching 14 tons. The CZ-5B is mainly used for heavy payloads in LEO, mostly for the launch of space station modules (Fig. 6.5). The CZ-5/YZ-2 is a three-and-a-half stage rocket, which can directly launch heavy satellites into GEO, medium Earth orbit, or SSO, with a GEO carrying capacity reaching 5.1 tons. The main technical parameters of the CZ-5 series rockets are shown in Table 6.1, and the carrying capacity is shown in Table 6.2.

2. Long March 9 Heavy-lift Launch Vehicle (CZ-9)

The Long March 9 launch vehicle is a new generation heavy-lift launch vehicle that China is currently developing. It will be used for future manned lunar landings, deep space exploration, and the construction of large-scale space infrastructure, with an expected first launch time around 2028. The Long March 9 heavy-lift launch vehicle will adopt a three-and-a-half stage configuration, divided into three configurations based on

Fig. 6.5 Long March 5 launch vehicle

different booster stages and the number of core stage engines: no booster configuration, two booster configuration, four booster configuration (Fig. 6.6).

The core stage has a diameter of 10 m and uses four 480 t-class liquid oxygen kerosene engines (five engines for the no booster configuration); the second stage uses two 220 t-class liquid oxygen liquid hydrogen engines; the third stage uses four 25 t-class liquid oxygen liquid hydrogen engines high-altitude improved type; the booster rocket has a diameter of 5 m and uses two 480 t-class liquid oxygen kerosene engines (Table 6.3). The current main technical parameters of the Long March 9 heavy-lift launch vehicle include:

Table 6.1 Main technical parameters of the long March 5 series launch vehicle

Category	Long March 5B (CZ.5B)	Long March 5 (CZ.5)	Long March 5/ Yuanzheng 2 (CZ.5/YZ.2)
Stages	1.5	2.5	3.5
Total length/m	53.66	56.97	
Width/m	17.3		
Takeoff thrust/t	1052 t		
Takeoff mass (excluding payload)/t	837	867	
Booster			
Length/m	27.6		
Diameter/m	3.35		
Engine	2 × YF.100		
Propellant	Liquid oxygen/kerosene		
Core stage one			
Stage length/m	33.16		
Diameter/m	5.0		
Engine	2 × YF.77		
Propellant	Liquid oxygen/liquid hydrogen		
Core stage two			
Stage length/m	No such structure	11.54	
Diameter/m		5.0	
Engine		2 × YF.75D	
Propellant		Liquid oxygen/liquid hydrogen	
Fairing			
Length/m	20.5	12.267	
Diameter/m	5.2	5.2	
Maximum payload envelope diameter/m	4.5	4.5	

- Rocket height: 103 m
- Maximum rocket width: 20 m
- Takeoff mass: 4137 t
- Takeoff thrust: 5873 t
- Maximum core stage diameter: 10 m
- Booster diameter: 5 m
- Number of boosters: 4
- Fairing diameter: 7.5 m (can be increased according to payload requirements)
- Fairing length: 25 m (can be increased according to payload requirements)
- LEO payload capacity: 140 t
- GTO payload capacity: 66 t

Table 6.2 Main payload capacity of long March 5 series launch vehicles

Launch vehicle	Target orbit	Orbit altitude/km	Orbit inclination/°	Payload capacity/t
Long March 5B (CZ.5B)	LEO	200 × 400	42	25
Long March 5 (CZ.5)	GTO	200 × 36,000	19.5	14
	TLI	200 × 380,000	24.5	8.2
	SSO	700 × 700	98	15
	MTO	200 × 26,000	55	13
	TMI	200 × 55,000,000		5
Long March 5/Expedition 2 (CZ.5/YZ.2)	GEO	36,000 × 36,000	0	5.1
	SSO	2000 × 2000	108	6.7
	MEO	26,000 × 26,000	55	4.5

LEO low earth orbit, *GTO* geostationary transfer orbit, *TLI* trans-lunar injection, *SSO* sun-synchronous orbit, *MTO* medium Earth transfer orbit, *TMI* trans-mars injection, *GEO* geosynchronous orbit, *MEO* medium earth orbit

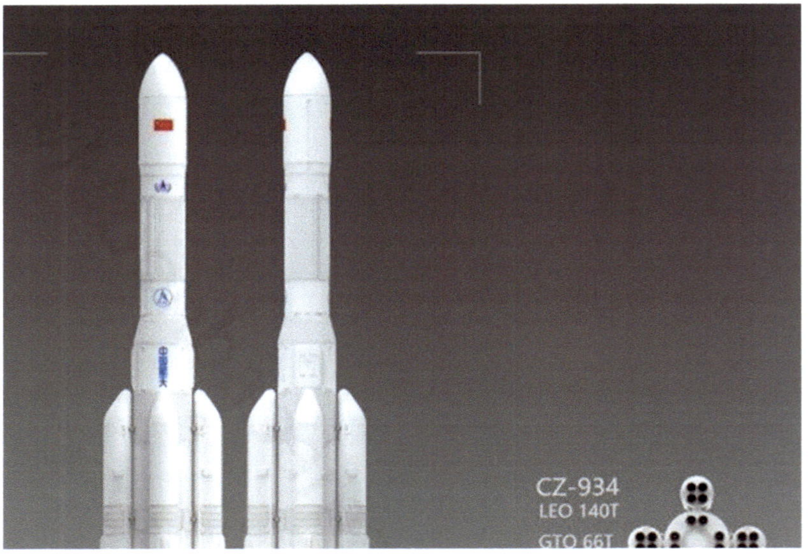

Fig. 6.6 Schematic diagram of long March 9 heavy launch vehicle configuration

- LTO payload capacity: 50 t
- Launch site: Wenchang Spacecraft Launch Site, China

3. Falcon Heavy Launch Vehicle

The Falcon Heavy launch vehicle, developed by SpaceX, is the most powerful operational launch vehicle in the world, with partial reusability (Fig. 6.7). On April 11, 2019, the Falcon Heavy successfully launched the "Arabsat-6A" communication satellite into its planned orbit and for the first time successfully recovered two booster rockets and the core stage rocket (Fig. 6.8). The Falcon Heavy is based on the Falcon 9, with two Falcon 9 first stages strapped on as booster rockets. Therefore, the length, fairing, and second-stage engine of the Falcon Heavy are identical to those of the Falcon 9, and it is mainly used for launching heavy satellites. The main parameters of the Falcon Heavy include:

- Rocket height: 70 m
- Maximum rocket width: 12.2 m
- Lift-off thrust: 2280 t
- Lift-off mass: 1420.8 t
- Fairing diameter: 5.2 m
- Fairing height: 13.2 m
- Maximum payload diameter: 4.6 m
- Maximum payload height: 11 m
- LEO payload capacity: 63.8 t
- GTO payload capacity: 26.7 t

Table 6.3 Technical parameters of LEO 140 t-class CZ-9 rocket

	Booster	Core stage one	Core stage two	Core stage three
Diameter/m	5	10	10 (liquid hydrogen tank)	7.5 (liquid hydrogen tank)
			7.5 (liquid oxygen tank)	5 (liquid oxygen tank)
Oxidizer/ propellant	Liquid oxygen/kerosene	Liquid oxygen/kerosene	Liquid oxygen/liquid hydrogen	Liquid oxygen/ liquid hydrogen
Engine	2480 t-class high-pressure combustion liquid oxygen kerosene engines	4480 t-class high-pressure combustion liquid oxygen kerosene engines	2220 t-class high-pressure combustion hydrogen oxygen engines	425 t-class expansion cycle hydrogen oxygen engines

Fig. 6.7 Falcon heavy launch vehicle. (Credit: Space X)

The Falcon Heavy launch vehicle is currently the rocket with the lowest launch price per unit mass internationally, with a single launch quote of 90 million US dollars. The corresponding LEO payload transport price is about 1500 US dollars/t, mainly due to its ability to recover three first-stage rockets and fairings, which can be reused after certain maintenance and testing. At present, the reuse record of Falcon 9 has reached 17 times, greatly reducing the cost of the launch vehicle.

4. Space Launch System

The Space Launch System (SLS) is a new generation of heavy-lift launch vehicle being developed by the United States, mainly used for Low Earth orbit and manned deep space exploration missions. The project was launched in 2011, inheriting the technical foundation of the Space Shuttle, Delta 4

Fig. 6.8 Recovery of Falcon heavy launch vehicle first-stage rocket. (Credit: Space X)

Heavy rocket, and Saturn V rocket. The rocket successfully launched for the first time in 2022 (Fig. 6.9). The SLS rocket has designed three types, namely SLS-1, SLS-1B, and SLS-2, which include manned and cargo types. SLS-1 uses a five-segment solid rocket booster, a universal core stage, and a transitional cryogenic upper stage, which can achieve a 70 t Low Earth orbit payload capacity and a 26 t lunar orbit payload capacity. SLS-1B uses the RL10-C3 hydrogen-oxygen engine's 8.4 m diameter upper stage, with a LEO payload capacity of about 95 t and a lunar orbit payload capacity of about 37 t. SLS-2 is based on SLS-1B, using advanced boosters to replace the five-segment solid rocket boosters, which can achieve a LEO payload capacity of about 130 t and a lunar orbit payload capacity of about 45 t. It is expected to be completed in 2028. The main design parameters of the current Space Launch System include:

- Rocket height: 111.25 m (SLS-2 cargo type)
- Number of rocket stages: 2
- LEO payload capacity:
  - SLS-1: 70 t
  - SLS-1B: 95 t
  - SLS-2: 130 t
- Lunar orbit payload capacity:
  - SLS-1: 26 t
  - SLS-1B: 37 t
  - SLS-2: 45 t
- Fairing diameter:
  - SLS-1B: 8.4 m
  - SLS-2: 10 m
- Launch site: John F. Kennedy Space Center

5. Starship System

In 2017, SpaceX released the design of the Falcon Heavy launch system with greater payload capacity, mainly for future large-scale satellite launches, manned lunar exploration, manned Mars exploration, and rapid intercontinental transportation, etc. It has now

Fig. 6.9 SLS heavy launch vehicle. (Credit: NASA)

officially been named the Starship system, a fully reusable heavy-lift rocket that will greatly increase payload capacity and reduce transportation costs. The entire system includes two stages, the first stage is the SuperHeavy, and the second stage is the Starship, both of which are fully reusable (Figs. 6.10). Currently, the main parameters of the Starship system include:

- Height: 120 m
- Diameter: 9 m
- Takeoff mass: 5000 t
- Stages: two stages
- Fairing diameter: 9 m
- Maximum payload height: 18 m
- LEO payload capacity (recovery): 150 t
- GTO payload capacity (recovery): 21 t
- GTO payload capacity (on-orbit refueling): 150 t
- Lunar and Mars payload capacity (on-orbit refueling): 100 t

The main parameters of the SuperHeavy include:

- Height: 70 m
- Diameter: 9 m
- Total mass: 3680 t
- Engine: 33 Raptor engines (cryogenic liquid oxygen methane)
- Specific impulse (sea level version): 334 s
- Thrust: 7590 t
- Direct return to the launch tower

The main parameters of the Starship class include:

- Height: 50 m
- Diameter: 9 m
- Dry mass: 120 t
- Cargo bay volume: 1000 m^3
- Propellant mass: 1200 t
- Engines: 6 Raptor engines (3 sea-level versions, 3 vacuum versions)
- Thrust: 1500 t
- Specific impulse (vacuum version): 382 s

The Starship class is divided into different configurations according to mission requirements, currently considering three types of missions:

1. As a spacecraft: Carrying multiple astronauts or space passengers, or carrying cargo to the moon, Mars, or other places on Earth.
2. As an interstellar refueling station: Loading the propellant needed by spacecraft, refueling spacecraft in Earth orbit, making manned

Fig. 6.10 Starship system. (Credit: Space X)

interstellar exploration and large-scale cargo transportation possible.

3. As a satellite deployment spacecraft: Equivalent to a cargo bay that can load multiple satellites, after entering orbit, deploy multiple spacecraft according to needs, can also be used for spacecraft recovery and space debris cleaning.

6.2.2 LEO-GEO Inter-Orbit Transporters

The transportation of SPSs requires the support of the LEO-GEO inter-orbit transporters. According to the different propulsion systems, there are three possible methods, as shown in Fig. 6.11. One is to use a high-thrust transport system (generally chemical propulsion) to directly transport to GEO, which needs to accelerate once in LEO to make the apogee reach GEO altitude, and then accelerate once in GEO to raise the perigee to the GEO altitude. The second is to use a medium-thrust transport system, which needs to gradually raise the apogee to the GEO altitude through multiple accelerations in LEO, and then gradually raise the perigee to the GEO altitude through multiple accelerations in GEO. The third is to use a continuous low-thrust transport system (generally electric propulsion), gradually accelerating to raise the orbit altitude in a spiral form to GEO.

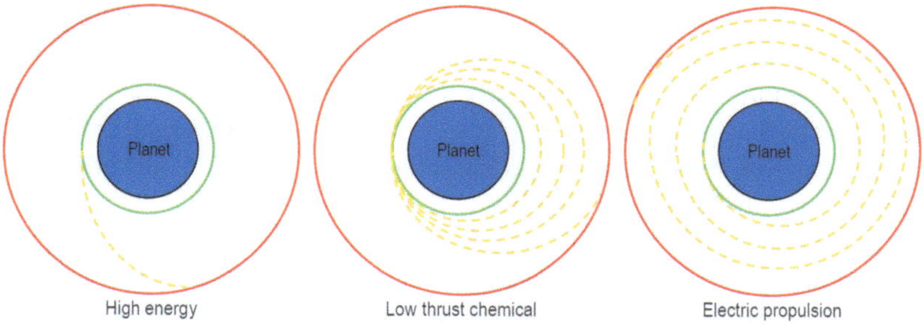

Fig. 6.11 Three LEO-GEO orbital transfer methods

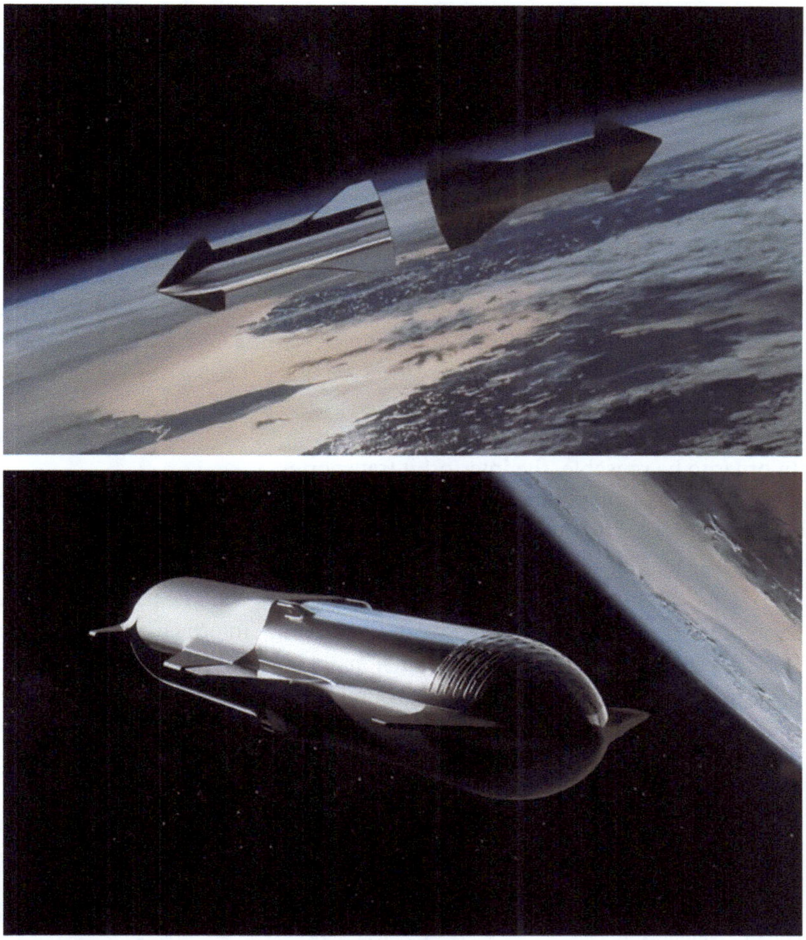

Fig. 6.12 Schematic of on-orbit refueling starship. (Credit: Space X)

The mass of the SPS is huge, and the use of a high-thrust transport system requires a large amount of thrust and propellant. At present, there is no practical LEO-GEO orbital transport system. The Starship with on-orbit refueling (Fig. 6.12) may become an important high-thrust orbital transport method. High-power electric propulsion for orbital transport is currently mainly considering internationally. Below is a concept of a reusable solar electric propulsion inter-orbit transporter.

The reusable solar electric propulsion inter-orbit transporter is used to capture and load SPS modules in LEO, and transport the modules to GEO using its own solar electric propulsion system, and deploy the SPS modules according to requirements, then return to low Earth orbit for the next transport. The reusable solar electric propulsion inter-orbit transporter has the following functions:

1. Function of autonomous payload capture

The inter-orbit transporter first needs to rendezvous with the payload, then use a robotic arm to grab the payload, and install the payload into the payload bay of the inter-orbit transporter through the operation of the robotic arm.

2. Function of orbit transfer

The inter-orbit transporter will rely on the solar electric propulsion system to transport the SPS modules to GEO, release the payload, and return to LEO in a short time.

3. Function of precise payload deployment

After the inter-orbit transporter arrives at the GEO, it needs to rendezvous with the space assembly service platform, use the robotic arm to release the payload, and precisely deploy it to the designated position on the platform.

4. Function of on-orbit refueling

In order to be reusable, the inter-orbit transporter needs to have the function of on-orbit propellant refueling, including propellant refueling in LEO and GEO.

A conceptual reusable solar electric propulsion inter-orbit transporter is shown in Fig. 6.13.

The reusable solar electric propulsion inter-orbit transporter corresponds to the ground-LEO transport capacity, which is about 100 t. In order to reduce the impact of long-term through the radiation belt during inter-orbit transport, a high-thrust electric propulsion system is needed. The main technical specifications are as follows:

1. Orbital transfer range: LEO-GEO-LEO
2. LEO altitude: 300 km
3. LEO inclination: 0°
4. Solar array output power: greater than 1 MW
5. Solar array output voltage: 500 V
6. Electric thruster: reference NASA-457Mv2
7. Number of electric thrusters: 20
8. Thrust: 46 N
9. Thruster specific impulse: 2740 s
10. Mass (excluding propellant and payload): 12 t. Specifically, it can be divided into:
    (a) Solar array: 3 t
    (b) Payload bay: 3 t
    (c) Service module: 1 t
    (d) Robotic arm: 1 t
    (e) Propulsion module: 4 t (excluding propellant). Specifically, it can be divided into:
        - Thruster: 2 t
        - Direct drive electric power supply: 1 t
        - Structure and thermal control: 1 t

The overall configuration includes a payload bay, service module, and propulsion module, which can be launched by a single launch vehicle. The payload bay mainly consists of a payload interface, robotic arm, and protective shield. The payload interface is used for installing the transported payload. The protective shield should envelop the entire payload, with a diameter of 11 m and a length of 25 m, to reduce the impact

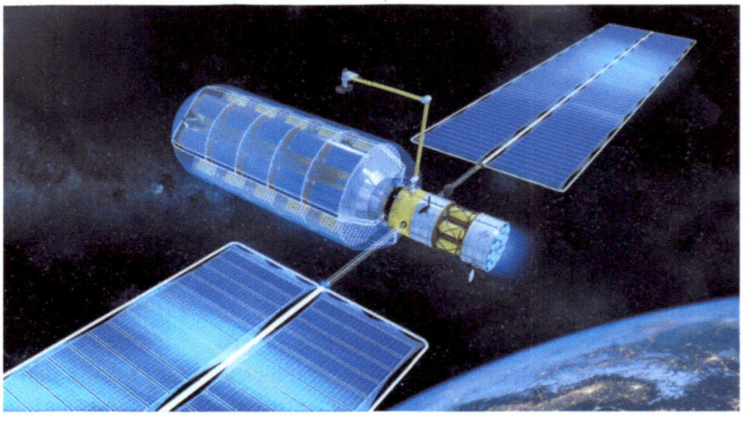

Fig. 6.13 Schematic diagram of a reusable solar electric propulsion inter-orbit transporter. (Credit: Xinbin Hou)

of radiation, debris, and thermal environment on the payload during orbital transfer. The robotic arm has a maximum operating distance of about 20 m, mainly used for payload capture, installation, and deployment. The service module has a diameter of 4 m and a height of 3 m, mainly composed of various service system equipment installed on the entire transporter, including solar arrays, power management and distribution, thermal management, attitude and orbit control management, telemetry and control, and integrated management equipment of the entire transporter. The propulsion module has a diameter of 4 m and a height of 5 m, mainly equipped with multiple high-power electric thrusters, propellant tanks, electric power supply, and management equipment.

The solar array consists of two sets of high-efficiency, high-power thin-film solar arrays, maintaining sun-orientation to achieve maximum power output. Each set of thin-film solar arrays includes two roll-out solar arrays, outputting electric power through a conductive rotary joint. The entire system includes four roll-out solar arrays and two high-power conductive rotary joints. A single roll-out solar array measures 10 m × 70 m, with an area of 700 m^2. The solar cells use multi-junction thin-film GaAs cells, considering a 40% solar cell efficiency and 80% filling rate, the power generation is about 300 kW. The entire solar array has an area of 2800 m^2, with a total power generation of about 1.2 MW, using a 600 V high voltage output to achieve direct drive electric power supply for the electric thrusters.

The electric propulsion subsystem includes high-power electric thrusters, on-orbit refueling propellant tanks, electric propulsion direct drive power supply units and electric propulsion management equipment. Considering the thrust demand of the inter-orbit transporter, the electric thruster is based on the highest power NASA-457 Mv^2(Fig. 6.14), which corresponds to a thrust of 2.3 N and a specific impulse of 2740 s under 50 kW power and 500 V operating voltage. According to the thrust demand, 20 50 kW Hall thrusters are installed at the end of the electric propulsion module(Fig. 6.15), and the surrounding thrusters have thrust direction steering capability. The front of the electric propulsion module is equipped with six cylindrical propellant tanks, which can carry up to 30 t of propellant and can be refueling in orbit.

Fig. 6.14 NASA.457Mv2 Electric thruster. (Credit: NASA)

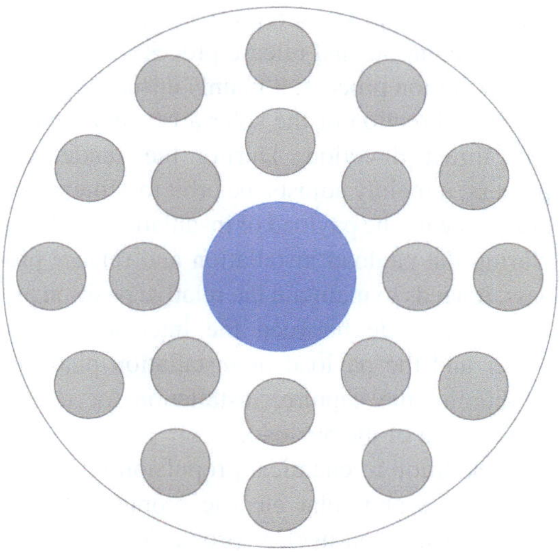

Fig. 6.15 Docking port and electric thruster layout

The structural subsystem mainly includes the payload interface and protective shield of the payload bay (for payloads that need space protection). The structural part of the service module and the electric propulsion module are used to install various equipment of the inter-orbit transporter to ensure the overall strength and rigidity. The payload interface is used to dock and install the transported payload under the operation of the robotic arm. The protective shield needs to be launched separately and deployed in orbit after entering orbit and then assembled with the inter-orbit transporter. The protective shield is opened during payload installation and release and is closed during transportation. The robotic arm, which is installed near the payload bay, is used to grab the payload and install it in the payload bay of the inter-orbit transporter. After reaching GEO and rendezvousing with the assembly service platform, the payload is deployed to the installation platform using the robotic arm. The AOC subsystem is primarily used for the attitude control and orbit maintenance of the inter-orbit transporter during the transportation process, rendezvous process, payload installation, and release process. During the transportation process, it mainly ensures the sun-oriented direction of the solar array and the correct thrust direction. During the rendezvous process, it mainly adjusts the orbit to achieve rendezvous with the payload or installation platform. During the payload installation and release process, it needs to maintain the relative position and relative attitude between the inter-orbit transporter and the payload or installation platform, facilitating the capture, installation, or release deployment of the payload.

In addition to chemical propulsion inter-orbit transporters and solar electric propulsion inter-orbit transporters, nuclear propulsion inter-orbit transporters are also one of the possible methods, as shown in Fig. 6.16, its electric power generation system uses a space nuclear reactor power source. Another unique inter-orbit transportation method is called the space elevator, as shown in Fig. 6.17, which lays a transportation cable between the ground and GEO orbit, maintains balance through the centrifugal force of the weight above the space elevator. The transporter relies on the energy transmitted by lasers as power to achieve round-trip transportation from the ground to space. Due to the high requirements for high-strength lightweight cables, there are currently no suitable materials, and this project is still in the conceptual stage.

Fig. 6.16 Nuclear propulsion inter-orbit transporter. (Credit: USNC Tech)

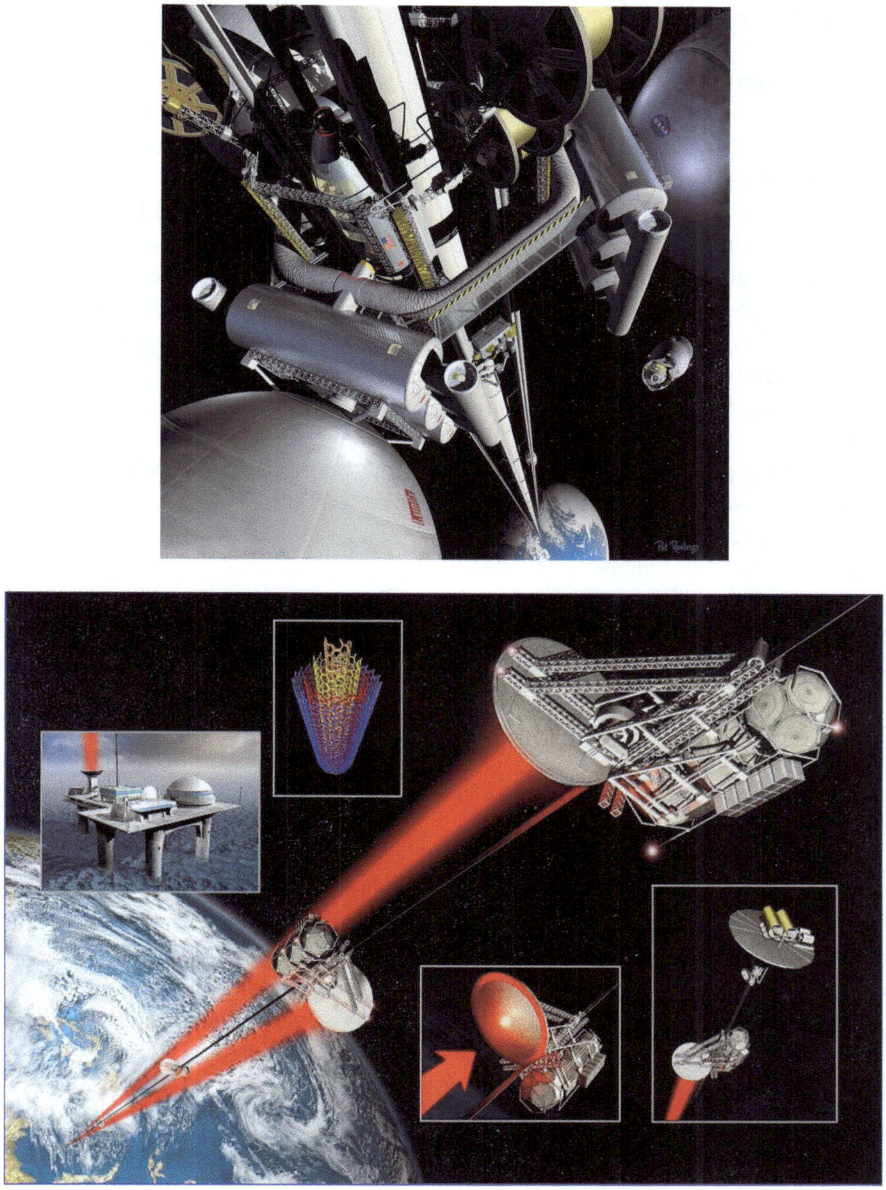

Fig. 6.17 Space elevator transport system. (Credit: NASA)

6.3 Assembly of SPSs

6.3.1 Requirements for Assembly Facilities

The SPS is a huge space system, with the following assembly characteristics.

1. Huge assembly scale

The SPS is huge in scale and mass, reaching several kilometers in length and nearly 10,000 t in total mass, composed of thousands of large modules, the number of assembly interface operations will reach thousands, the assembly scale is huge.

2. Complex assembly process

The structure of the SPS is complex, involving different assembly modules and various assembly positions. The number of assembly interfaces is huge. The assembly planning, module transportation trajectory, robot motion trajectory, and assembly operations are extremely complex, requiring multi-robot cooperative assembly.

3. Complex assembly environment

The assembly location of the SPS is located in the geostationary orbit and can only be assembled by robots. The space environment is harsh, including radiation environment, high and low temperature environment, and complex sunlight conditions. Assembly and maintenance may involve tens of kilovolts of high voltage and microwave radiation with a power intensity reaching thousands of watts per square meter.

4. High assembly efficiency requirements

The assembly construction time of the SPS should be limited to 1–2 years. Completing the complex assembly operations of thousands of modules within a few hundred days puts extremely high requirements on the efficiency of assembly.

The on-orbit assembly facilities of the SPS need to include a space assembly service platform and various space assembly robots. Figure 6.18 shows a schematic diagram of the microwave transmitting antenna robot assembly.

6.3.2 Space Assembly Service Platform

The space assembly service platform is mainly used for the temporary placement of assembly modules, docking maintenance of assembly robots, propellant refueling of inter-orbit transporters and assembly robots, etc., and the following factors need to be considered.

1. Operational capability: Capable of capturing assembly modules of the inter-orbit transporter from a considerable distance and placing them on the space assembly service platform; capable of separating assembly modules for easy capture and transport by assembly robots.
2. Loading capability: Capable of loading all assembly modules transported by the inter-orbit transporter.
3. Environmental protection of assembly modules: Provides necessary environmental protection for temporarily placed assembly

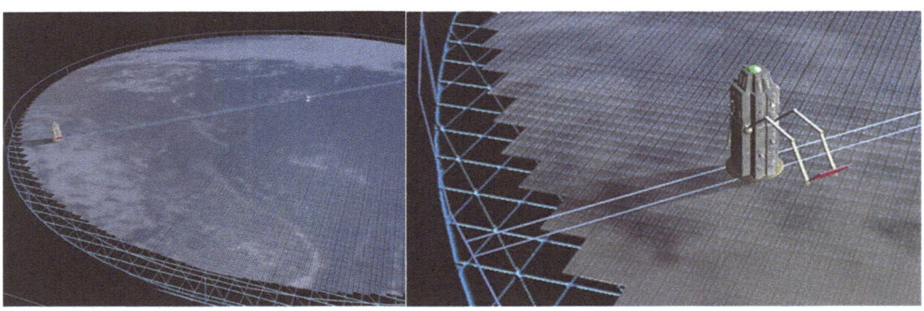

Fig. 6.18 Schematic diagram of microwave transmitting antenna robot assembly

modules, such as maintaining a reasonable storage temperature.

4. Docking and maintenance capability of assembly robots: Provides a docking platform for assembly robots, enabling them to refueling propellant, energy, perform maintenance, and provide long-term accommodation and environmental protection for robots.
5. Propellant storage capability: Capable of storing enough propellant required for inter-orbit transporters to return from GEO to LEO and can achieve long-term storage.
6. Propellant refueling capability: Capable of providing sufficient propellant supply for inter-orbit transporters returning to LEO; capable of providing propellant supply for assembly robots.
7. Orbit maintenance and maneuverability: Capable of maintaining a specific orbit and adjusting the orbital position according to assembly requirements.

The technical features of the space assembly service platform envisioned for the SPSs are as follows.

1. Operating orbit: Geostationary orbit, located near the orbit of the assembled SPS, adjusting the orbital position according to the expansion of the scale of the SPS during the assembly process.
2. Scale of the space assembly service platform gradually expands with the assembly requirements of the SPS.
3. Equipped with a large robotic arm, used to unload all modules transported by the inter-orbit transporter, and place them in a fixed position on the space assembly service platform.
4. Use the robotic arm to capture individual modules and to place in a specific position on the space assembly service platform which is convenient for the capture and transport of space assembly robots.
5. Equipped with propellant storage tanks and on-orbit propellant refueling platforms, supporting the propellant supply of inter-orbit transporters through the robotic arm.
6. Equipped with multiple robot-docking platforms, used to simultaneously support the docking of multiple space assembly robots, realizing the propellant supply, charging, and module capture of assembly robots.
7. Equipped with a solar power supply system, attitude orbit maintenance system, thermal control system, and telemetry communication system, realizing platform orbit adjustment, telemetry communication, equipment power supply, temperature control, etc.
8. Technical parameters of the space assembly service platform:
   (a) Length: Approximately 50 m.
   (b) Length after placing two transport packages: Approximately 100 m.
   (c) Width: Approximately 10 m.
   (d) Capable of simultaneously accepting the rendezvous of two inter-orbit transporters, able to accommodate all modules transported by two inter-orbit transporters, using electromagnetic docking mechanisms.
   (e) Can support the docking of 10 space assembly robots simultaneously.
   (f) Propellant storage capacity: 50 t.
   (g) Power supply capacity: 500 kW.
   (h) Robotic arm operation capability: Maximum unloading mass is 100 t and maximum operating range is 30 m.
   (i) Propellant resupply distance for inter-orbit transporter: 30 m.

The service platform for space assembly is still in the conceptual research stage. The United States proposed a concept of space logistics base for the construction of Solar Power Satellite in 2007, as shown in Fig. 6.19.

6.3.3 Space Assembly Robots

The space assembly robots have the functions of grabbing assembly modules, transporting modules to specific locations, assembling modules, and assisting in the deployment of modules. They mainly includes platform mobile robots and free maneuvering robots. Considering the assembly process, transporting the assembly module from the assembly service platform to the assembly location is mainly completed by the free maneu-

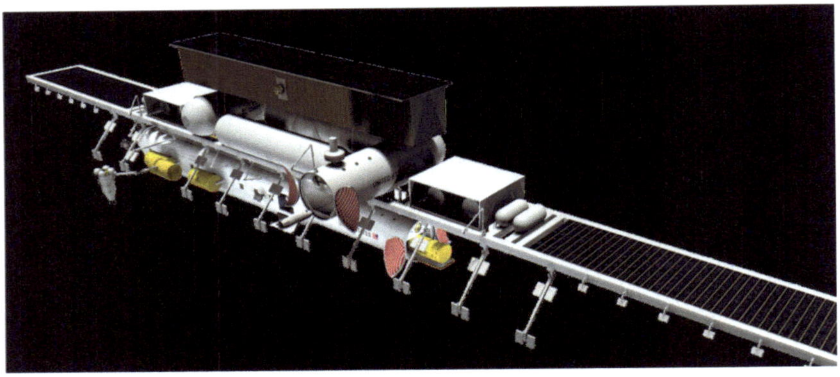

Fig. 6.19 Concept of space logistics base. (Credit: James Michael Snead)

vering robot. For small modules, a single free maneuvering robot can complete the transportation, and for large modules, multiple free maneuvering robots are needed to complete the transportation. For modules transported by a single free maneuvering robot, when approaching the docking target location, the assembly module is passed to the platform mobile robot, which completes the docking and assembly of the module. For large modules that need to be transported by multiple free maneuvering robots, when approaching the docking target location, the assembly module is passed to the platform mobile robot, which together with other free maneuvering robots, completes the docking and assembly of the module. Considering typical assembly modules and assembly states, the following factors need to be considered for space assembly robots.

1. Mobility and maneuverability: The assembly robot's range of movement and maneuverability need to be determined based on the assembly and deployment requirements of the assembly module. For platform mobile robots, the movement speed and the requirements of the attached platform need to be considered. For free maneuvering robots, the robot's maneuverability, maneuvering range, and movement speed need to be considered. Mobility or maneuverability directly determines the size of the assembly objects.
2. Operational capability: The carrying capacity of the assembly robot, the operation scale and space of the robotic arm, and the dexterity of the robotic hand determine the weight, size, and assembly interface status of the assembly objects.
3. Autonomy: To improve assembly efficiency and reduce ground support for space assembly, space assembly robots need to have the ability to autonomously plan movement trajectories, move autonomously, autonomously grasp targets, and assemble autonomously without direct human participation.
4. Collaborative ability: The size of the Solar Power Satellite is huge, and the corresponding assembly module size is also very large, requiring multiple assembly robots to operate together to assemble complex large modules.
5. Energy supply method: The entire process of the space assembly robot's operation requires power support. Considering that the range of movement of the assembly robot is large, it can be considered to use a combination of self-contained power system, wired power supply, and wireless power supply, including body-mounted solar cells, batteries, wired power interfaces, and wireless power devices (including close-range electromagnetic induction power supply and long-distance laser wireless power supply, etc.).
6. Propellant resupply: Free maneuvering robots need to consume propellant during their maneuvering process in space. In order to achieve long-term operation of free maneuvering robots, they need to have the ability to use space assembly service platforms for on-orbit propellant refueling.

7. Space environment adaptability: Space assembly robots are in the unprotected space including the harsh high-energy particle radiation environment, sunlight environment, high and low temperature environment, and possible high voltage and high-power intensity microwave radiation environment. They can operate reliably for a long time.
8. Low cost: Through the mass production, autonomy, high-efficiency and long-life operation of space assembly robots, the cost of space assembly process can be reduced.

6.3.3.1 Platform Mobile Robots

Platform mobile robots move and complete space assembly and maintenance operations through a moving platform on a truss or other structure or relying on their own mobile system. Platform mobile robots do not need a complex orbit maneuvering system, do not need to consume propellant, and the moving range can be infinitely extended with the expansion of the structure.

The mechanical arm operation system of the International Space Station is a typical platform mobile robot. Almost all large structures of the space station are assembled through the mechanical arm operation system, which is also the core equipment for astronauts to carry out on-orbit assembly and maintenance operation. It mainly consists of the Space Station Remote Manipulator System (SSRMS), Special Purpose Dexterous Manipulator (SPDM), and Mobile Base System (MBS) (Figs. 6.20, 6.21, and 6.22).

The SSRMS is a seven-degree-of-freedom mechanical arm, 17.6 m long, weighing 1800 kg, with a maximum load capacity of up to 100 t, peak power of 2 kW, and average power of 1360 W. The SSRMS can be combined with the SPDM for more precise assembly operation. The SPDM consists of two flexible mechanical arms, the ends of which can be fitted with payloads and various space operation tools, and can be used for the installation and removal of equipment. The main technical parameters of the SPDM are as follows: length 3.5 m, width 0.88 m, weight 1662 kg, maximum load 600 kg, 15 degrees of freedom, maximum power 2000 W, average power 600 W.

The SSRMS can move on the space station in two ways. First, the two ends of the mechanical arm are equipped with Latching End Effectors (LEE), which can attach and lock a special mechanism on the space station—the Power Data Grapple Fixtures (PDGF), to provide power, data, and command support for the SSRMS. There are multiple PDGFs set up on the space station, and the mechanical arm can move autonomously within a certain range by attaching to PDGFs in different positions. In addition, the space station is equipped with a Mobile Base System (MBS) that can move on the truss. The MBS is installed on the rail, allowing free movement on the truss, and includes four PDGF interfaces, which can greatly expand the operating range of the space station remote operation system. The MBS measures 5.7 m × 4.5 m × 2.9 m, weighs 1450 kg, has

Fig. 6.20 Space station remote operation system. (Credit: Canadian Space Agency)

Fig. 6.21 End-specific dexterous hand. (Credit: Canadian Space Agency)

a maximum load of 20,900 kg, a maximum power of 825 W, and an average power of 365 W.

In the late 1990s, in response to the assembly needs of SPSs, Carnegie Mellon University in the United States cooperated with NASA to develop a space robot prototype called Skyworker (Fig. 6.23), which has the ability to move autonomously on the truss and autonomously transport payloads, for on-orbit assembly, inspection, and maintenance tasks. Its main assembly objects include solar arrays, microwave transmitting antennas, and power systems, and representative work includes:

Fig. 6.22 Space station mobile platform MBS. (Credit: Canadian Space Agency)

Fig. 6.23 Skyworker robot prototype. (Credit: Carnegie Mellon University)

1. Walking and turning on the truss structure, and crossing between trusses
2. Grabbing and placing payloads at any position and direction in space
3. Walking and turning with load, as well as crossing between trusses
4. Performing on-orbit inspection tasks
5. Connecting power transmission and signal transmission cables
6. Collaborative operation of multiple robots, transporting large payloads
7. Carrying out other tasks that require the cooperation of multiple robots

Skyworker includes 11 motion joints, using motor drive, planetary gears, and harmonic gear two-stage reducer methods. It has various sensors, force sensors are used to measure the force applied by Skyworker, joint angle sensors assist in completing tasks by measuring the rotation angles of structural joints and fixtures, and position sensors measure the direction and position of targets through infrared sensors installed on the fixtures.

6.3.3.2 Free Maneuvering Robot

The core of the free maneuvering robot is its ability to maneuver on the orbit, possess the ability to maneuver freely, and is equipped with a mechanical arm system that can grab payloads and complete corresponding assembly operations after reaching the target position. The free maneuvering robot is equivalent to a spacecraft equipped with a powerful maneuvering ability and an operating mechanical arm, so its operating range is greatly increased. It can achieve rapid transportation and assembly of payloads and is convenient for efficient assembly of ultra-large space structures. Currently, research on free maneuvering robots is in the initial stage, and there are no actual products available for large spacecraft assembly.

The Orbital Express program implemented by the United States in 2007 provided some verification of the functions of free maneuvering robots. The Orbital Express program includes two spacecraft, one is the service spacecraft (ASTRO), and the other is the serviced spacecraft (CSC) (Fig. 6.24). The main technologies verified include target approach, position holding, target capture, docking, fuel supply, on-orbit detection, repair, etc. The ASTRO spacecraft is equivalent to a free maneuvering robot, consisting of six parts, namely the satellite platform, docking mechanical arm system, rendezvous approach sensor, capture system, fuel supply system, and component replacement unit. The operating stages of Orbital Express include the approach stage, capture stage, docking stage, and service stage. Among them, the approach stage mainly includes fly-around, approach, and position holding. Capture uses two methods: direct capture docking method and mechanical arm capture method. The direct capture docking method is to gradually approach the serviced spacecraft through satellite maneuvering, and use the three-pronged docking mechanism to connect the two spacecraft together. The mechanical arm capture method is when the satellite maneuvers to a certain distance from the serviced spacecraft, the mechanical arm is used to grab the target, and the two spacecraft are connected together using the docking mechanism through the mechanical arm.

In February 2020, the United States successfully launched the Mission Extension Vehicle (MEV-1), as shown in Fig. 6.25, achieving on-orbit docking with the Intelsat-901 satellite nearing the end of its life, thereby replacing the propulsion system of the Intelsat-901 satellite, achieving the satellite's life extension, and providing a good verification of the on-orbit rendezvous and docking technology in the GEO orbit.

Fig. 6.24 Schematic diagram of orbital express. (Credit: NASA)

Fig. 6.25 Schematic diagram of MEV-1 satellite on-orbit service mission. (Credit: Northrop Grumman)

6.4 Decommission at the End of SPS's Life

After the mission of the SPS is over, it must be disposed of to prevent long-term occupation of the geostationary orbit or the generation of space debris affecting other spacecraft. The Inter-Agency Space Debris Coordination Committee (IADC) is an international organization coordinating space debris issues. In 2002, the IADC officially released the "IADC Space Debris Mitigation Guidelines," which became a programmatic document guiding countries around the world to effectively control the generation of space debris during space activities. It stipulates the requirements and mitigation measures that spacecraft and launch vehicles should take to reduce the generation of space debris and reduce the degree of harm to space systems. For the main operating orbits of spacecraft, the "IADC Space Debris Mitigation Guidelines" provides recommendations for orbit protection zones, as shown in Fig. 6.26.

1. LEO protection zone A, the entire spherical space from the surface of the earth to a height of 2000 km

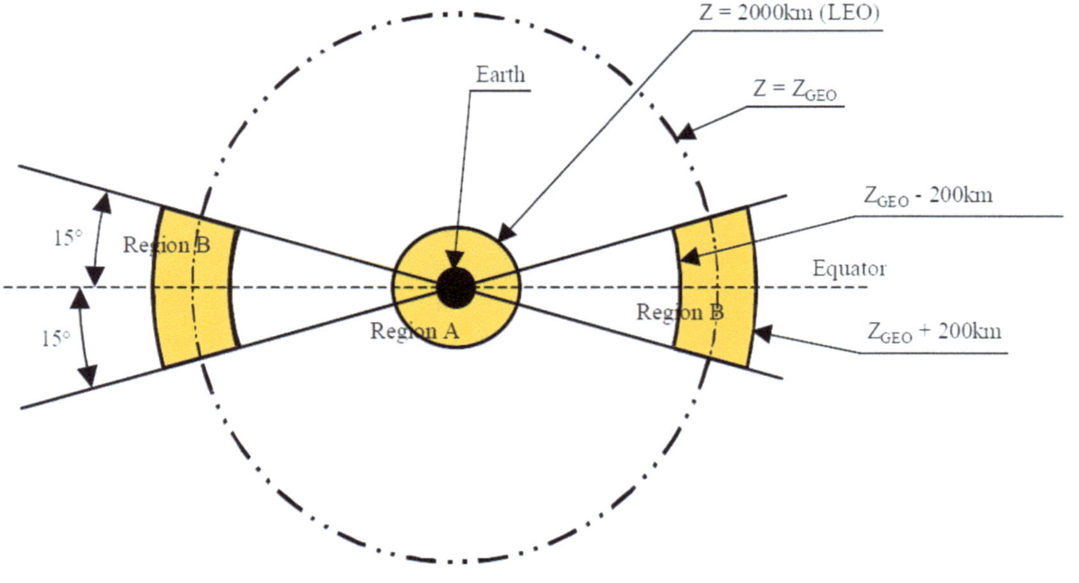

Fig. 6.26 IADC orbital protection area

2. GEO protection zone B, a part of the spherical shell space near the geostationary orbit, defined as follows:
   (a) GEO altitude: 35,786 km
   (b) Lower boundary of the zone altitude: 200 km below the GEO altitude
   (c) Upper boundary of the zone altitude: 200 km above the GEO altitude
   (d) Latitude range: ±15°

For LEO spacecraft that have completed their missions, deorbit operations should be carried out, preferably direct re-entry into the atmosphere or return to the ground, or enter orbits with a stay time not exceeding 25 years by orbital maneuvering. For GEO spacecraft that have completed their missions, they should maneuver to orbits higher than the GEO protection zone, with an eccentricity not greater than 0.003, and the minimum increment of the perigee height of the orbit relative to the geostationary orbit is:

$$235\,\text{km} + \left(1000\, C_R\, A / m\right) \tag{6.1}$$

where C_R is the solar radiation pressure coefficient and A/m is the area-to-mass ratio, m^2/kg.

The Solar Power Satellite is a large-sized GEO spacecraft. After the mission is over, it needs to be disposed of in accordance with the "IADC Space Debris Mitigation Guidelines" to meet the requirements of orbit debris mitigation. The following methods can be mainly considered:

1. Transport the entire SPS to an orbit area higher than the GEO protection zone. This method is relatively simple to operate, but it requires a large amount of propellant.
2. Transport the entire SPS to low Earth orbit for deorbit operations. This method requires more propellant due to its large volume and mass and there is a certain risk when re-entering the atmosphere.
3. Reutilization of SPS. The components of the SPS can be disassembled. The structure of the SPS is preserved for direct use in the construction of subsequent SPS. Other components are reused through in situ resource utilization and on-orbit manufacturing technologies. Parts that cannot be utilized are transported to orbital regions higher than the GEO protection zone by inter-orbit transporters, or transported to LEO for deorbit operations.

Bibliography

1. Smitherman D V. Space Elevators: An Advanced Earth-Space Infrastructure for the New Millennium[R]. NASA/CP-2000-210429. Marshall Space Flight Center, Huntsville, Alabama, 2000.
2. Edwards B C. The Space Elevator[R]. NIAC Phase II Final Report, 2003.
3. Inter-Agency Space Debris Coordination Committee. IADC Space Debris Mitigation Guideline, IADC-02-01[R]. IADC, 2002.
4. Inter-Agency Space Debris Coordination Committee. IADC Space Debris Mitigation Guidelines, Revision 1, IADC-02-01[R]. IADC, 2007.
5. Hou X B, Zhang X H, Wang L. Introduction on Space Solar Power[M] (in Chinese). Beijing: China Astronautic Publishing House, 2020.
6. Butow S, Cooley T, Felt E, et al. State of the Space Industrial Base 2020[R]. 2020.
7. Olson J, Butow S, Felt E, et al. State of the Space Industrial Base 2022[R]. 2022.
8. https://images.nasa.gov/
9. www.spacex.com

7 Multi-Rotary Joints Solar Power Satellite

7.1 Composition of MR-SPS

The Multi-Rotary joints Solar Power Satellite (MR-SPS) is a representative of non-concentration continuous transmission SPS. It uses solar photovoltaic power generation and microwave wireless power transmission methods. The power generated by solar arrays is transmitted to the microwave transmitting antenna through special configuration, greatly reducing the power of single conductive rotary joint (Figs. 7.1 and 7.2).

MR-SPS consists of the Solar Energy Collection and Conversion (SECC) subsystem, the Power Transmission and Management (PTM) subsystem, the Microwave Wireless Power Transmission (MWPT) subsystem, the Structure subsystem, the Attitude and Orbit Control (AOC) subsystem, the Thermal Control (TC) subsystem, and the Information and System Operation Management (ISOM) subsystem (Fig. 7.3). The main function of the SECC subsystem is to collect incident sunlight and convert solar energy into electrical power. The PTM subsystem transmits high-power electrical power to the MWPT subsystem and provides the required electric power for the equipment of other subsystems. The MWPT converts high-power electric energy into high-power microwave energy and transmits energy to the ground using a large-diameter transmitting antenna, is the core subsystem of the MR-SPS. The structural subsystem connects all subsystems together, provides necessary rigidity and strength, provides installation interfaces for SPS's equipment, and provides a platform for system maintenance. The AOC subsystem ensures accurate Sun orientation of the solar arrays and accurate Earth orientation of the microwave transmitting antenna which ensures the system safety and efficiency. The AOC subsystem also maintains the reasonable orbit position and altitude of the system. As a super high-power, super large-scale space system, thermal control devices are scattered everywhere which ensures the normal operating temperature of each subsystem's equipment and components. The ISOM subsystem is responsible for collecting the operating information of the entire SPS system and managing the operation of the entire system.

X. Hou, *Space Solar Power*, Sustainable Development Goals Series,
https://doi.org/10.1007/978-981-97-3580-8_7

Fig. 7.1 Schematic diagram of MR-SPS. (Credit: Xinbin Hou)

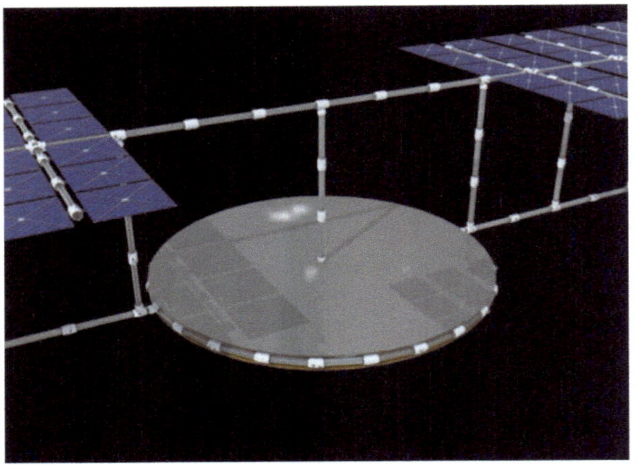

Fig. 7.2 Partial view of MR-SPS

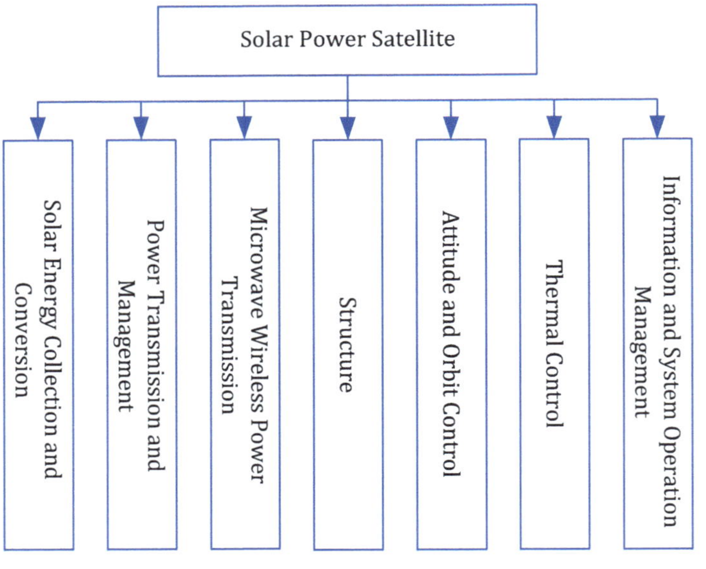

Fig. 7.3 Composition of MR-SPS

7.2 Configuration of MR-SPS

The overall configuration of the MR-SPS is shown in Fig. 7.4, mainly composed of three major parts: the solar arrays (south, north), the microwave transmitting antenna, and the main structure. The microwave transmitting antenna is located in the center and points to the ground. The solar arrays are located in the south and north directions and are sun-oriented through the conductive rotary joints. The main structure connects the solar arrays and the microwave transmitting antenna. PTM equipment, AOC equipment, ISOM equipment, TC equipment, etc. are installed on the structure frames of the solar arrays, microwave transmitting antenna, and main structure.

The solar arrays of a GW-level MR-SPS consist of 50 solar subarrays (25 each in the south and north), each subarray measures 200 m × 600 m, and the gap between the subarrays is 10 m. In order to avoid affecting the antenna, the solar subarrays are not arranged at the position corresponding to the antenna between the south and north solar arrays. Considering the impact of the ecliptic angle, about 150 m of distance is left empty at both ends of the antenna without arranging the solar subarray, and the total length of the solar array structure is about 11,800 m.

The frequency of transmitting microwave is chosen to be 5.8 GHz which considers the transmission distance of 36,000 km from space to the ground, as well as factors such as the size matching of the transmitting antenna and the receiving antenna and the power intensity. The diameter of the microwave transmitting antenna is set to be 1 km.

Considering mechanical performance, the microwave transmitting antenna is arranged in the middle, with solar arrays arranged on both sides, and the three parts are connected through the main structure. The MR-SPS adopts a special configuration, the entire main structure is composed of two north-south main truss structures and multiple up-down main truss structures. The north-south main truss structure above is used to support the solar subarrays. The multiple conductive rotary joints installed on the main truss are used to drive the solar subarrays to achieve independent rotation of each solar subarray. The north-south main truss structure below is mainly used to support the microwave transmitting antenna. The two north-south main truss structures are connected together by multiple up-down main truss structures to form the entire structure. Among them, the up-down main truss structures and the north-south main truss structure below are also used for the installation of power transmission cables.

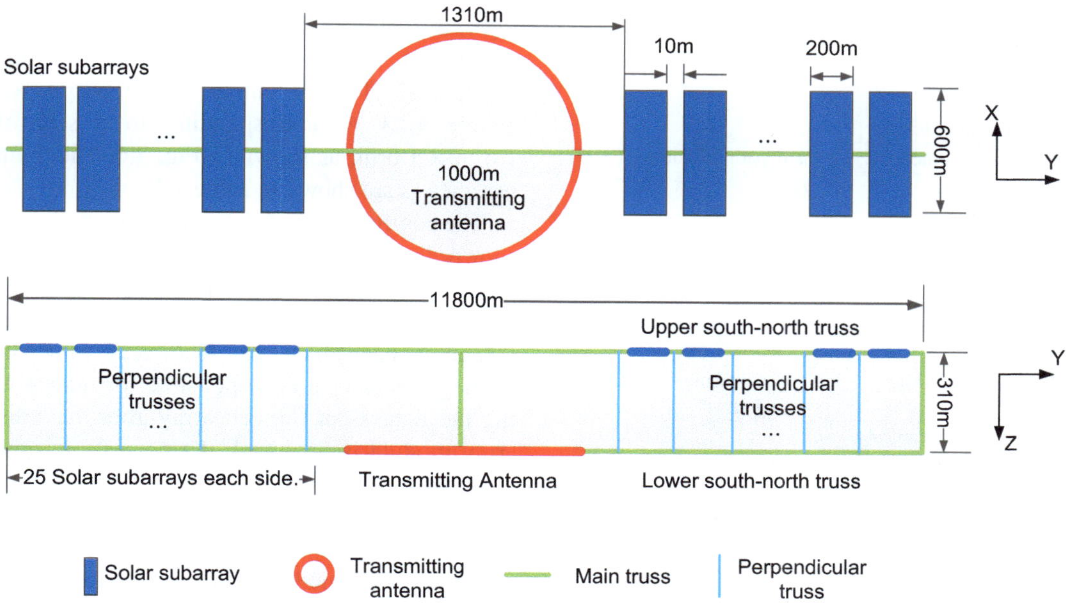

Fig. 7.4 Overall configuration of MR-SPS

7.3 Energy Conversion Efficiency

The energy efficiency chain of the MR-SPS is shown in Table 7.1.

Table 7.1 Energy efficiency chain of MR-SPS

Impact item	Efficiency	Total efficiency
SECC subsystem efficiency (0.29)		
Solar cell efficiency	0.40	0.4
Sun pointing efficiency	0.99	0.396
Solar array filling factor	0.85	0.336
Average efficiency of sunlight angle change	0.958 (23.44°)	0.322
Space environment factor	0.90	0.290
PTM subsystem efficiency (0.857)		
Subarray voltage conversion efficiency	0.97	0.281
Bus voltage conversion efficiency	0.98	0.276
Power transmission efficiency	0.95	0.262
Bus voltage step-down conversion efficiency	0.98	0.256
Microwave source voltage conversion efficiency	0.97	0.249
Service subsystem power consumption efficiency	0.999	0.2477
Microwave conversion and emitting efficiency (0.76)		
Microwave conversion efficiency	0.80	0.198
Microwave modulation efficiency	0.95	0.188
Microwave atmospheric transmission efficiency (0.95)		
Microwave transmission efficiency	0.95	0.179
Microwave reception and conversion efficiency on the ground (0.727)		
Beam collection efficiency	0.95	0.170
Antenna reception efficiency	0.9 (considering pointing error)	0.153
Rectifying efficiency	0.85	0.130
Power regulation efficiency (0.96)		
Power connecting efficiency	0.98	0.127
Power regulation efficiency	0.98	0.124

7.4 Preliminary Scheme of Primary Subsystems

7.4.1 Solar Energy Collection and Conversion Subsystem

The entire SECC includes 50 solar subarrays (25 each in the north and south directions). Each solar subarray includes 12 solar array modules, arranged in two columns, each column has 6 solar array modules, and the size of a single solar array module is 100 m × 100 m. Each solar subarray rotates to maintain sun orientation through two conductive rotary joints connected to the north-south main truss structure, and transmits electric power to the transmission cable on the main structure, and then transmits it to the microwave transmitting antenna (Fig. 7.5). The solar array module is the basic module of the SECC subsystem, which is launched into orbit and deployed on orbit, and is assembled to form the entire solar array.

Considering high efficiency and lightweight, thin-film multi-junction gallium arsenide (GaAs) solar cells are selected. Assuming the size of a thin-film GaAs cell unit is 0.06 m × 0.12 m and the conversion efficiency can reach 40% (AM0). Multiple cell units are arranged on a 15 μm Kapton film substrate to form a cell module, which is the basic unit for the folding and unfolding of the solar array module. A cell module uses 13 parallel and 26 series, and the output voltage is about 127.4 V, the output current is approximately 10.4 A, corresponding to a size of 1.6 m × 1.6 m, as shown in Fig. 7.6. The main parameters are shown in Table 7.2.

The configuration of the solar array module is shown in Fig. 7.7, including the deployment truss, thin-film cell array, array film tensioning mechanism, power transmission interface, assembly mechanism, etc. The array film tensioning mechanism is responsible for maintaining the flatness of the cell array after the solar array module is deployed, the power transmission interface is responsible for power output of the solar array module, and the assembly mechanism is used for the structural connection between the solar array module and the structure truss,

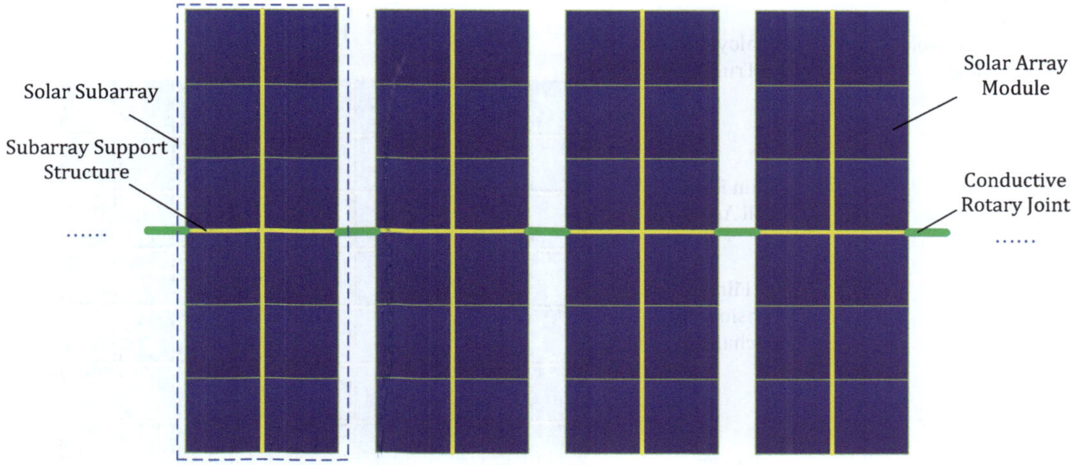

Fig. 7.5 Schematic diagram of the solar array

Fig. 7.6 Thin-film cell module

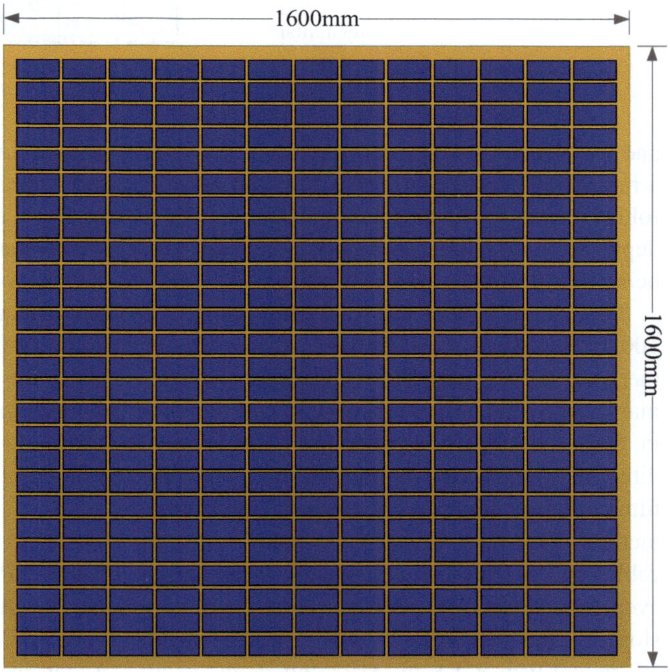

Table 7.2 Parameters of the cell module

Area S_0/m^2	Number of solar cells	Output voltage of optimal operating point/V	Output current of optimal operating point/A	Area density/(g/m^2)	Thickness/mm	Specific power/(W/kg)
2.56 (1.6 × 1.6)	338	127.4	10.4	180	0.05	2875

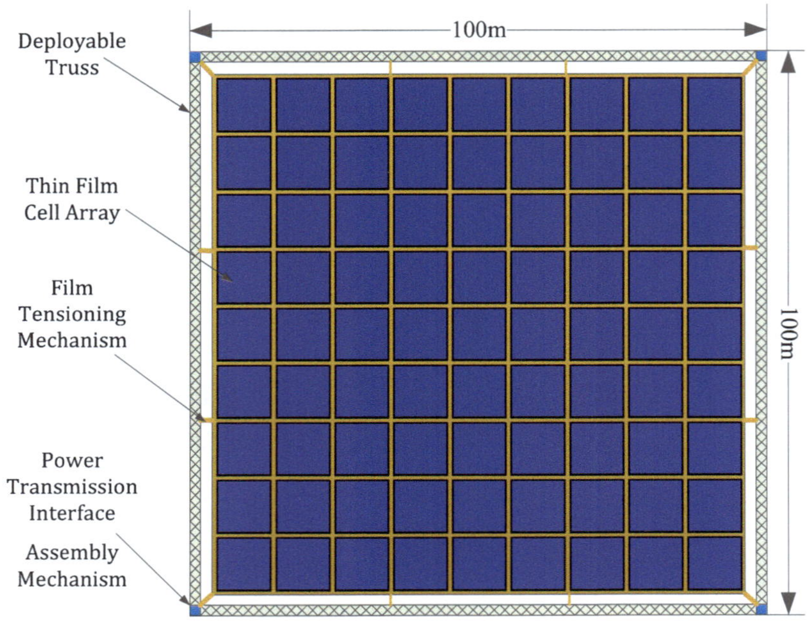

Fig. 7.7 Configuration scheme of the solar array module

and the structural connection between the solar array modules. The entire thin-film cell array is folded in two dimensions during launching and is deployed in two dimensions using the truss after being transported to the target orbit.

The overall size of the solar array module is 100 m × 100 m. Considering various factors, there are 57 cell modules in one direction (primary deployment direction) and 56 cell modules in the other direction (secondary deployment direction), with a total number of 3192. The thin-film cell array adopts a two-dimensional folding method (Fig. 7.8). Considering factors such as cables and folding bending radius, assuming an average thickness of 0.5 mm per layer, the dimensions after folding are approximately 1.9 m × 1.9 m × 1.8 m. 3192 cell modules are installed on a thin-film substrate with space radiation protection function. The upper surface needs to be covered with a radiation protection layer with high transmittance to sunlight. The outer surfaces of the upper and lower surfaces are conductive layers to prevent the occurrence of space discharge, as shown in Fig. 7.9. It is assumed that the thickness of the packaging thin-film cell array is 0.25 mm.

The solar array module uses a two-dimensional truss as the support structure (Fig. 7.10), with a truss arranged on each of the four sides in the *X* and *Y* directions. The truss and the thin-film cell array are tensioned by the array surface tensioning mechanism. Truss drive devices and docking devices between the solar array modules are arranged at the four corners in the *X* and *Y* directions. The *Z* direction is the packing structure of the entire solar array module. Considering the margin, the maximum envelope size of the solar array module in the folding state is 4 m × 4 m × 2 m (Fig. 7.11).

7.4.2 Power Transmission and Management Subsystem

According to the configuration characteristics of the MR-SPS, power transmission and management adopt a distributed + centralized method. First, the power generated by the solar subarray is converged to the solar subarray bus. The subarray power is collected through two conductive rotary joint output cables and the cables of the main truss structure in the up-down and north-south directions, and transmitted to the microwave transmitting antenna, as shown in Fig. 7.12. The entire system on the solar array uses different voltage level supply buses, including 500 V

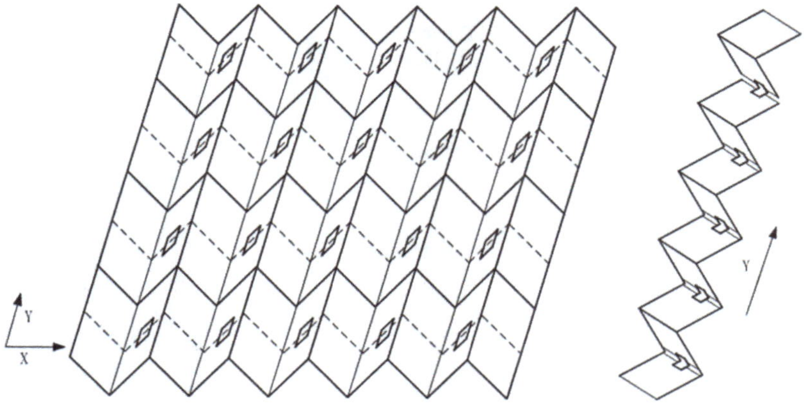

Fig. 7.8 Two-dimensional folding of the thin-film cell array

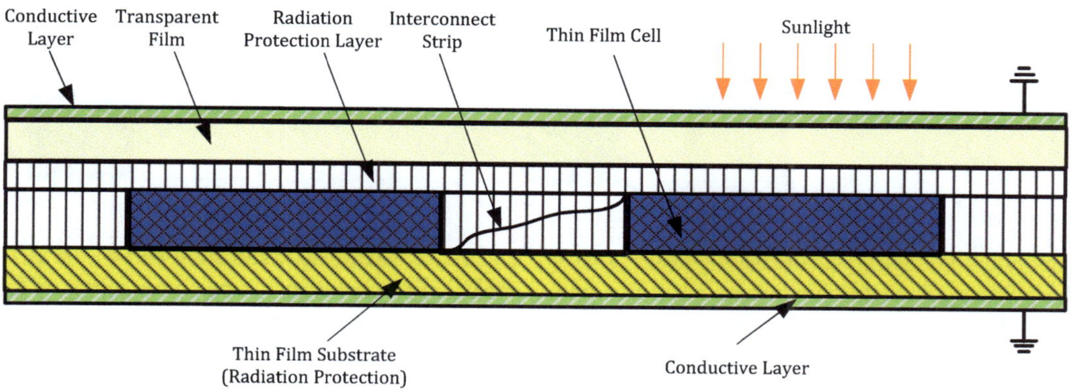

Fig. 7.9 Packaging structure of the flexible thin-film cell array

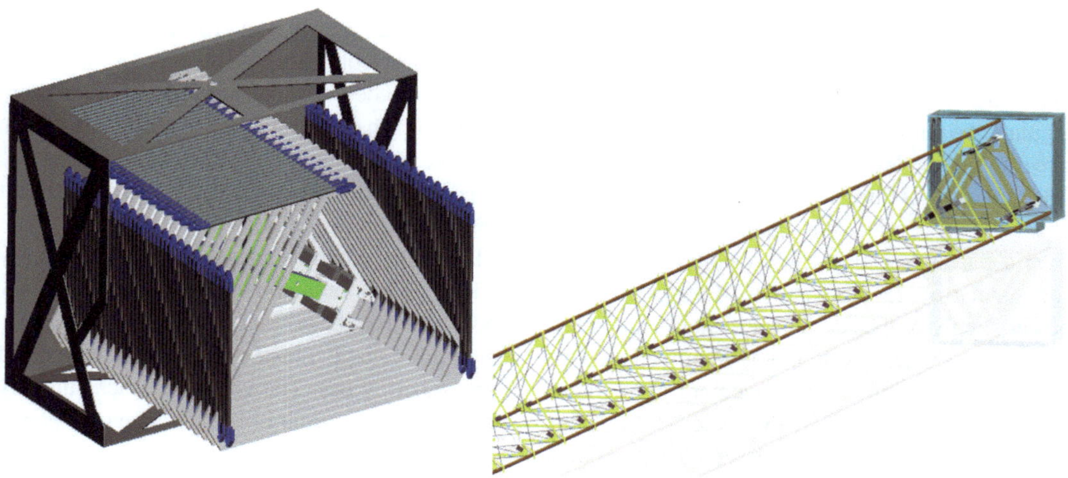

Fig. 7.10 Folding and unfolding of a solar array module truss

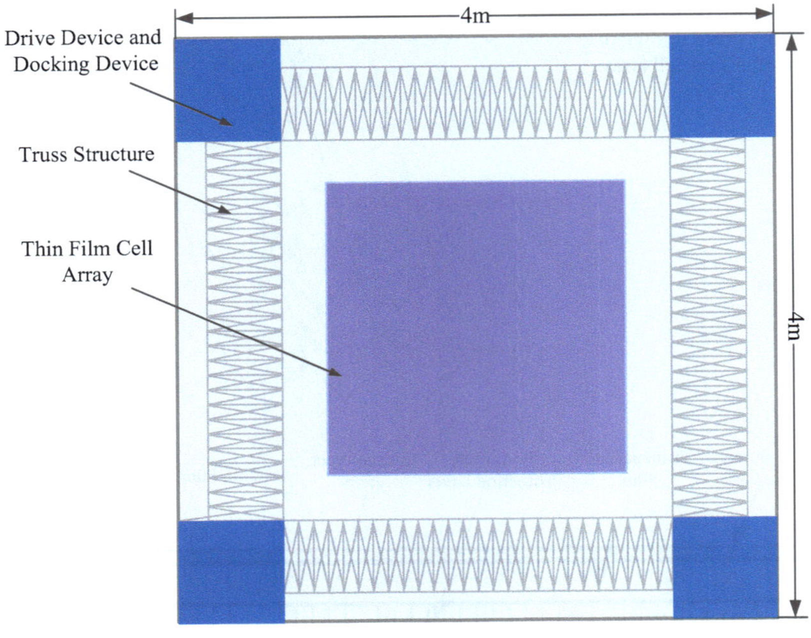

Fig. 7.11 Scheme of folding a solar array module

medium voltage transmission bus, 5 kV high-voltage transmission bus, and 100 V supply bus. The main structure transmission bus is 20 kV, and the microwave transmitting antenna includes 5 kV high-voltage bus, 500 V medium voltage bus. The service subsystem bus includes 5 kV high voltage, 500 V medium voltage, and 100 V low voltage.

The main function of power transmission and management in the solar array area is to transform the voltage of the power from the solar subarray, regulate the current, and transmit it to the conductive rotary joint. Part of the power generated by the solar subarray is directly distributed for the power supply of the service subsystem equipment on the solar subarray, and batteries are configured on the solar subarray to store electric energy for the power supply of the service subsystem equipment during the shadow period. The main function of power transmission and management in the main structure area is to regulate the power output from the solar subarray through the conductive rotary joints, use the power transmission bus on the main structure to transmit the power to the transmitting antenna, distribute part of the power for the power supply of the related service subsystem equipment installed on the structure (including high-power electric thrusters, etc.), and store part of the electric energy by batteries for the power supply of the service subsystem equipment during the shadow period. The main function of power transmission and management in the transmitting antenna array area is to distribute the power transmitted by the power transmission bus, transform the voltage, transmit it to the power supply unit of the microwave source of the microwave transmitting antenna, distribute part of the power for the power supply of the service subsystem equipment installed on the microwave transmitting antenna, and store part of the energy by batteries for the power supply of the service subsystem equipment during the shadow period.

The solar subarray consists of 12 solar array modules. The output power of each solar array module needs to be boosted to 5 kV through a subarray high-voltage transformer unit and connected to the subarray bus. The voltage of each subarray bus is 5 kV; the current is 4800 A. The total power is about 24 MW and is connected to the corresponding conductive rotary joint. Each solar subarray needs to configure a subarray service bus for the power supply of the service subsystem equipment using 100 V low-voltage bus (Fig. 7.13). At the same time, each solar array module is equipped

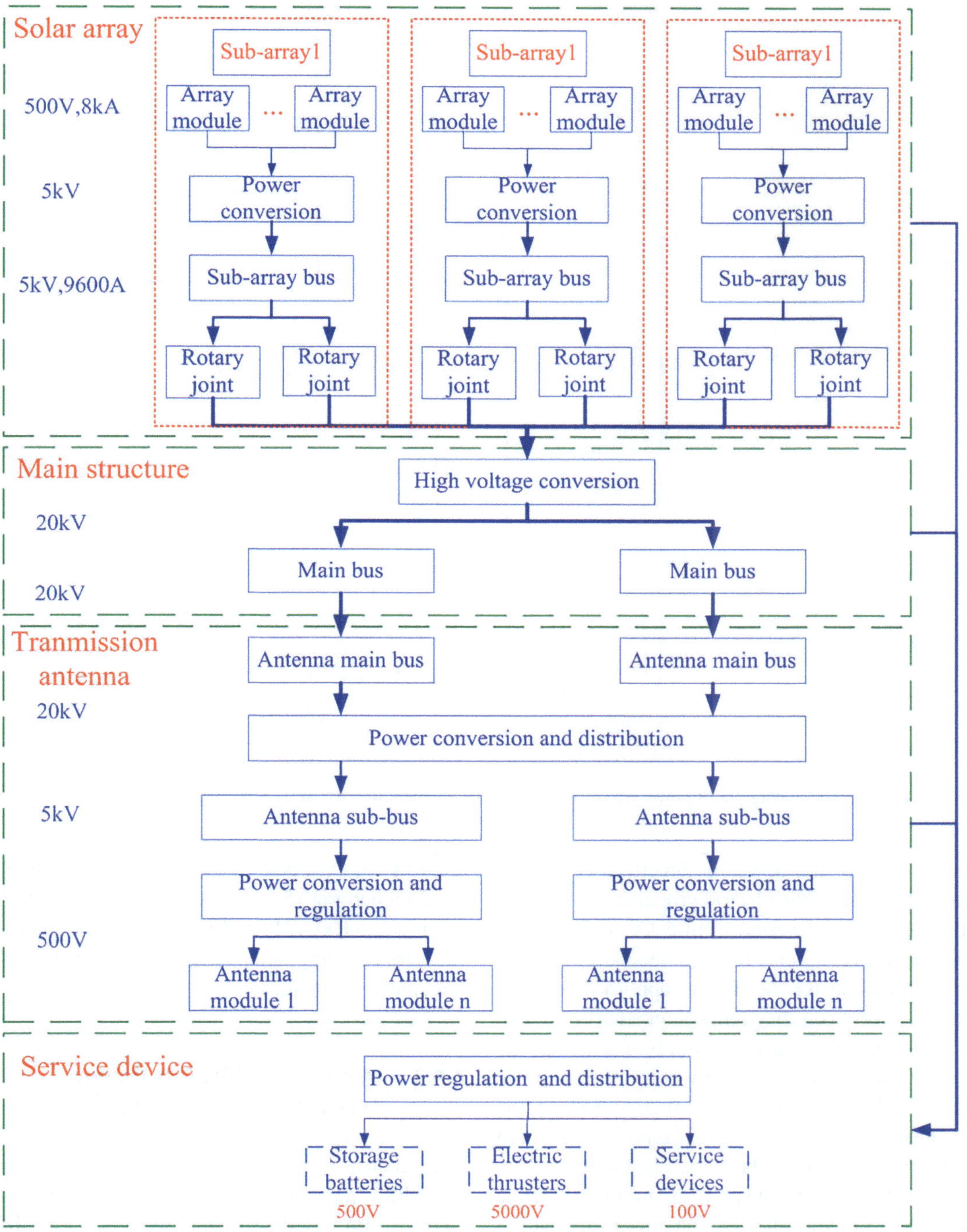

Fig. 7.12 Schematic diagram of distributed + centralized power transmission and management

with a energy storage device to ensure the normal operation of basic service equipment during shadow periods and fault periods.

The power of 100 subarray buses corresponding to 50 solar subarrays passes through the conductive rotary joints and the main power will be boosted to 20 kV through the bus high-voltage transformer units, connected to the power transmission buses, and connected to the microwave transmitting antenna through the two interfaces in the north and

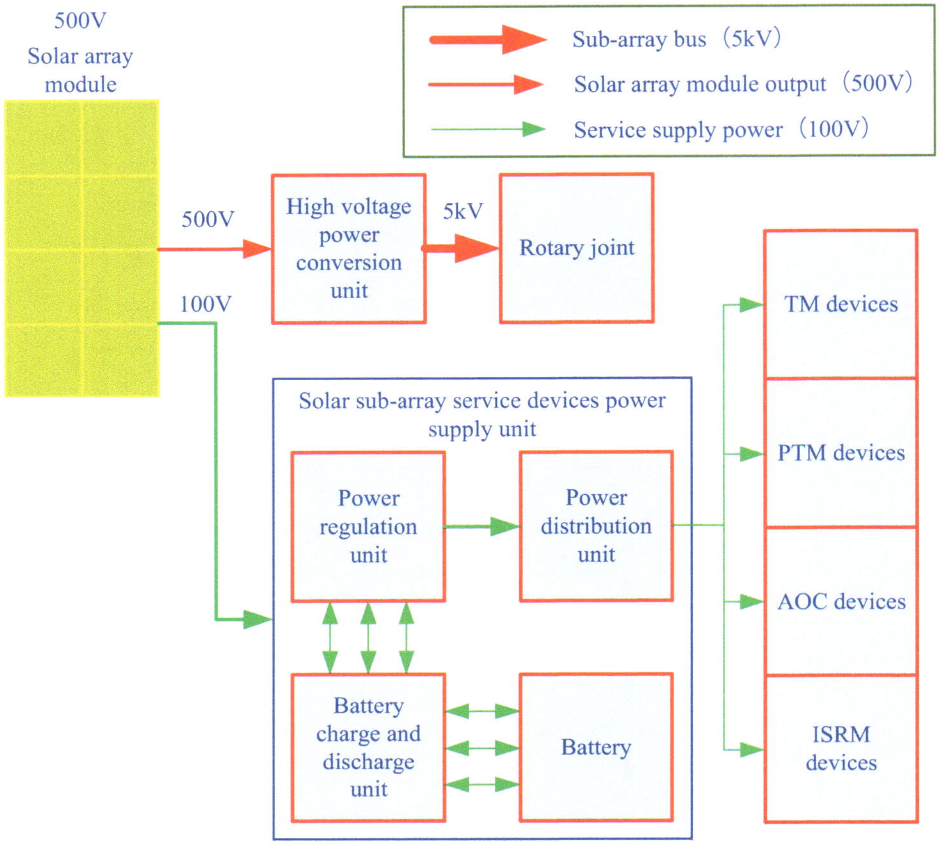

Fig. 7.13 Schematic diagram of power transmission and management of solar array module

south. Part of the power is directly used for the power supply of the electric propulsion system, or the voltage is converted from 5 kV to 500 V for battery charging, and through secondary voltage transformation for the power supply of the main structure service subsystem equipment (Fig. 7.14).

For the microwave transmitting antenna, the power transmitted by the power transmission bus needs to be first converted from 20 to 5 kV by the voltage step-down transformer unit, and output to multiple antenna sub-busbars. Each antenna sub-bus distributes the main power to multiple antenna assembly modules through the power distribution units, and then converts the voltage from 5 kV to 500 V through the voltage step-down transformer units, and supplies power for the antenna subarrays through multiple output ports. Part of the power of the antenna sub-bus is directly used for the power supply of the electric propulsion system (5 kV). Or it is used for the power supply of battery charging (500 V) and service subsystem equipment (100 V/28 V/12 V/5 V) of the microwave transmitting antenna array by various voltage conversion (Fig. 7.15).

7.4.3 Microwave Wireless Power Transmission Subsystem

The microwave transmitting antenna of the MR-SPS is a flat array antenna with a diameter of 1 km. To simplify the structure of the antenna, the circular microwave transmitting antenna is simplified into an octagonal structure, composed of 80 100 m × 100 m antenna structure subarrays, as shown in Fig. 7.16. Each antenna structure subarray (Fig. 7.17) includes five 20 m × 100 m antenna assembly modules (Fig. 7.18). The antenna assem-

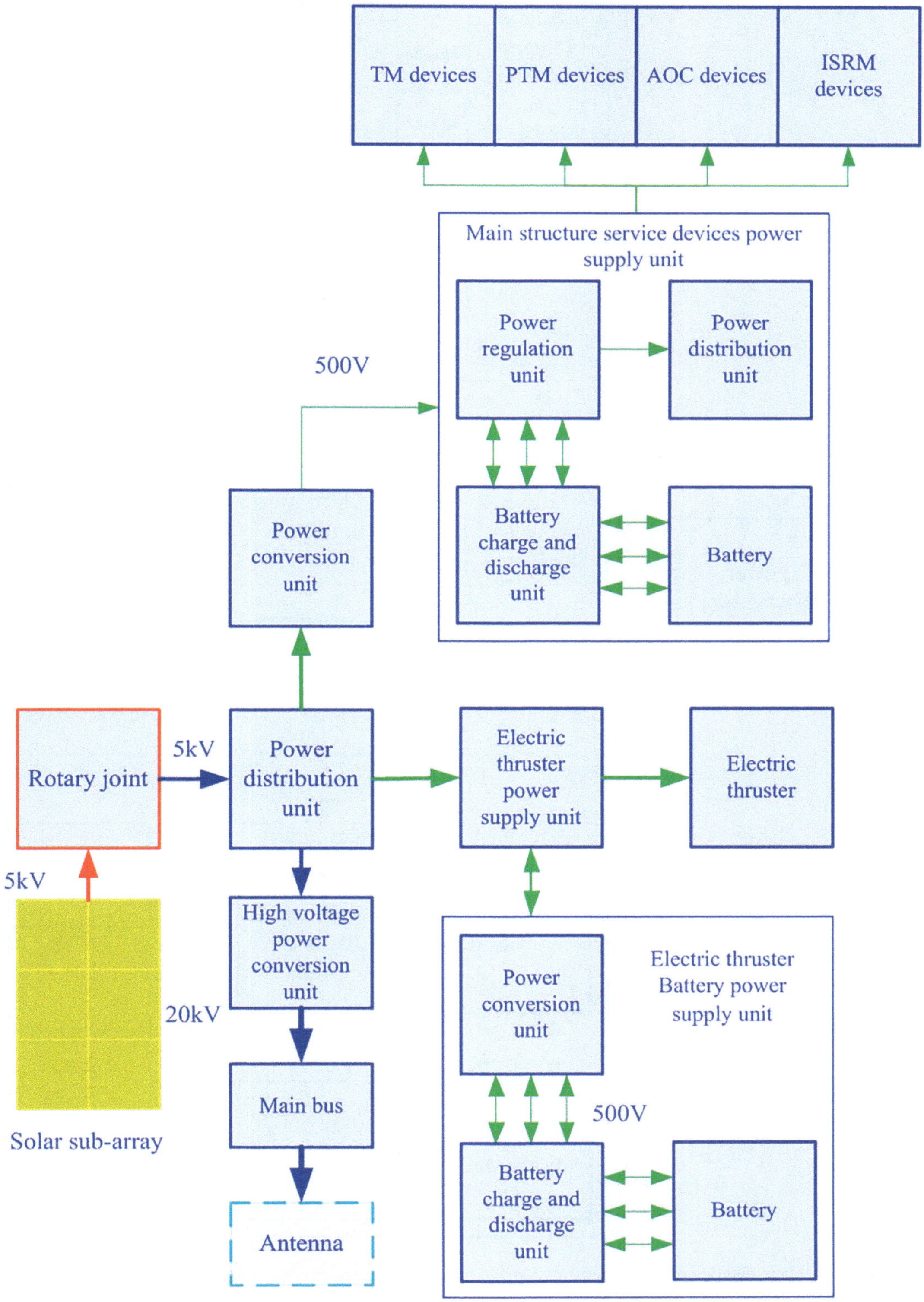

Fig. 7.14 Schematic diagram of power transmission and management of main structure

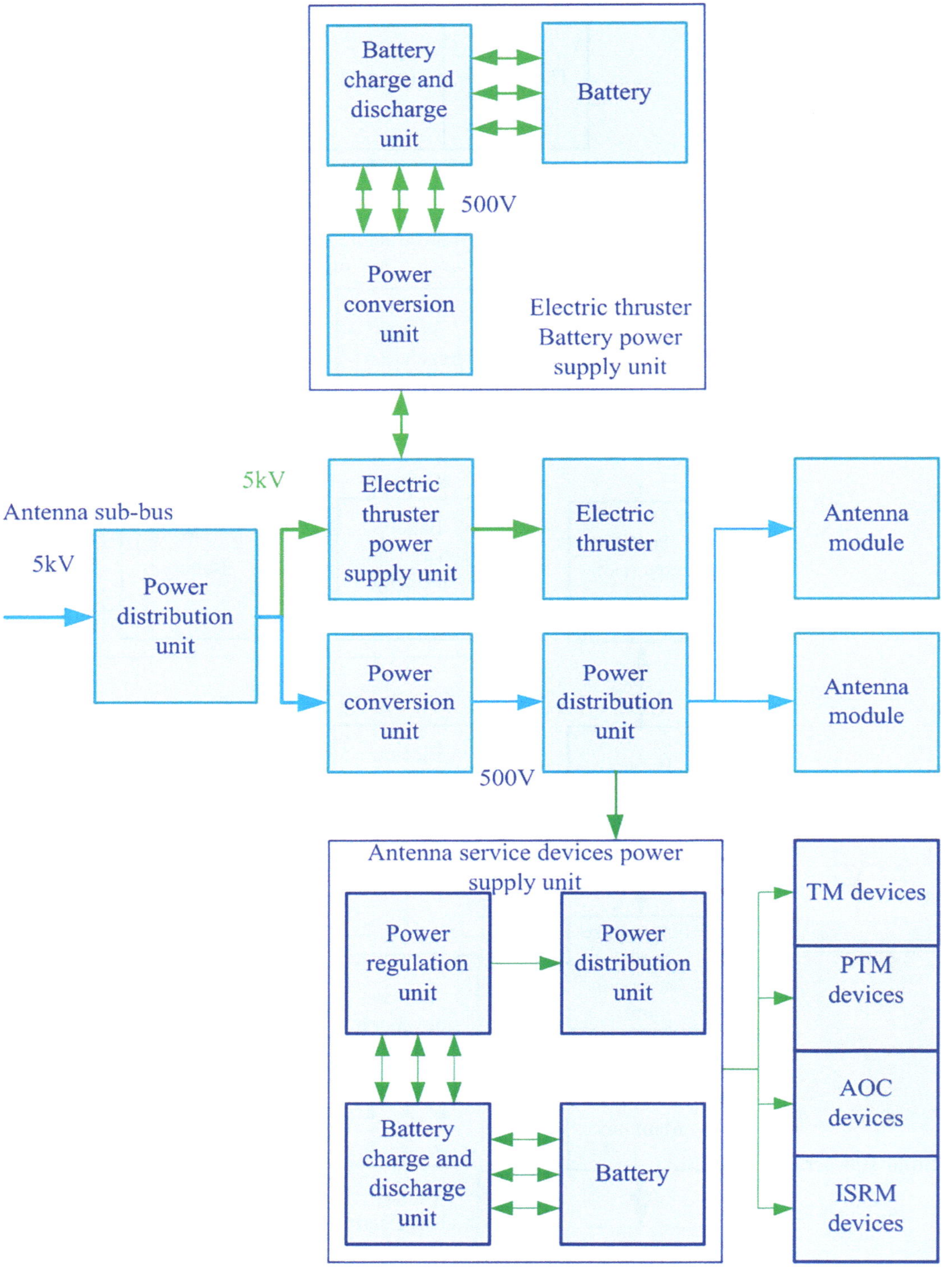

Fig. 7.15 Schematic diagram of power transmission and management at microwave antenna end

Fig. 7.16 Schematic diagram of microwave transmitting antenna configuration

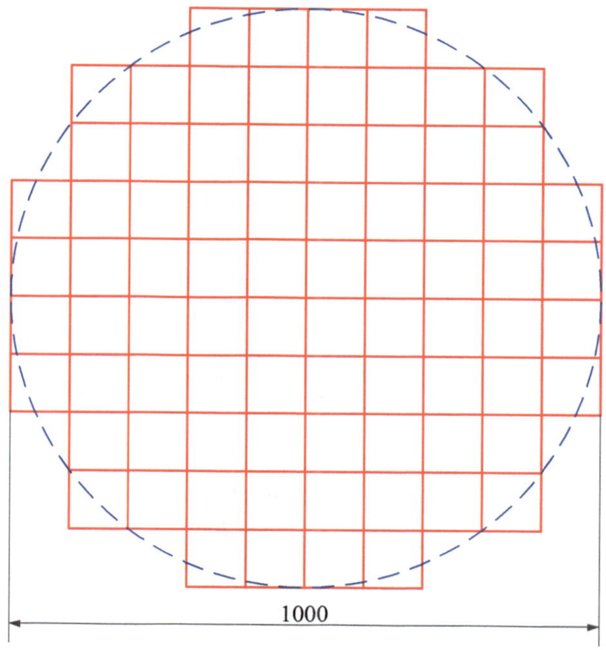

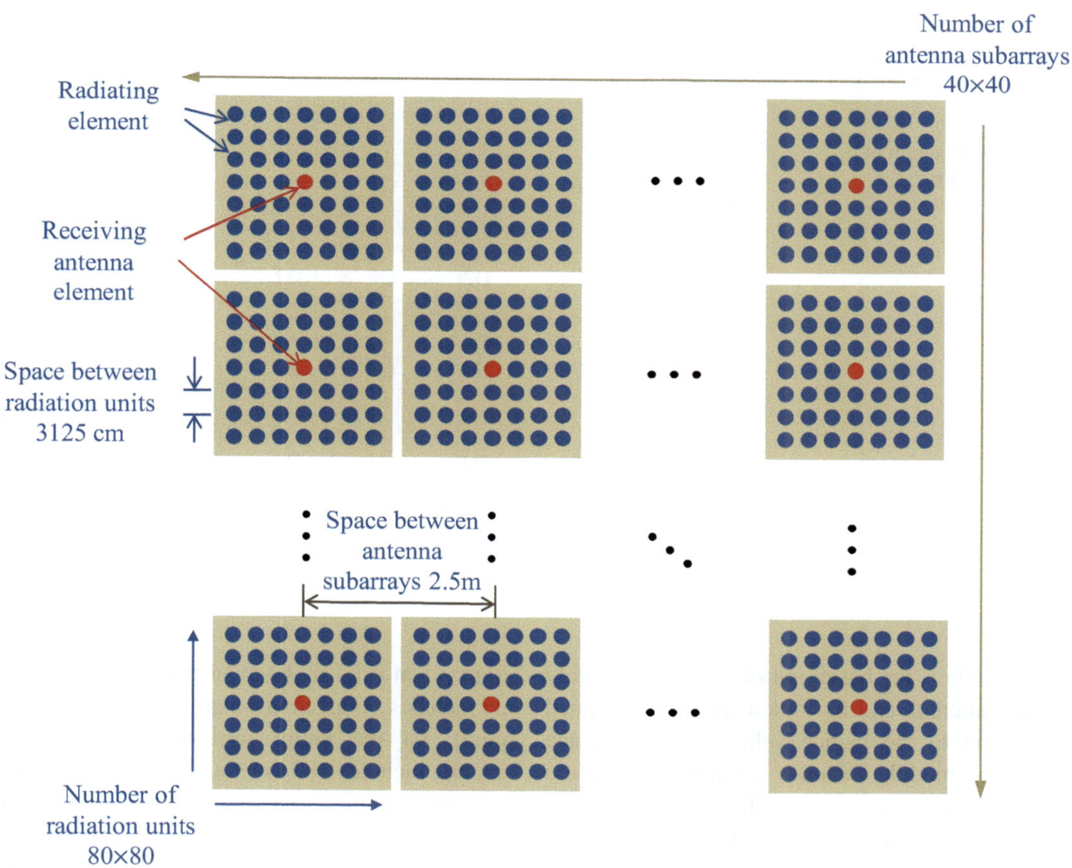

Fig. 7.17 Schematic diagram of antenna structure subarray

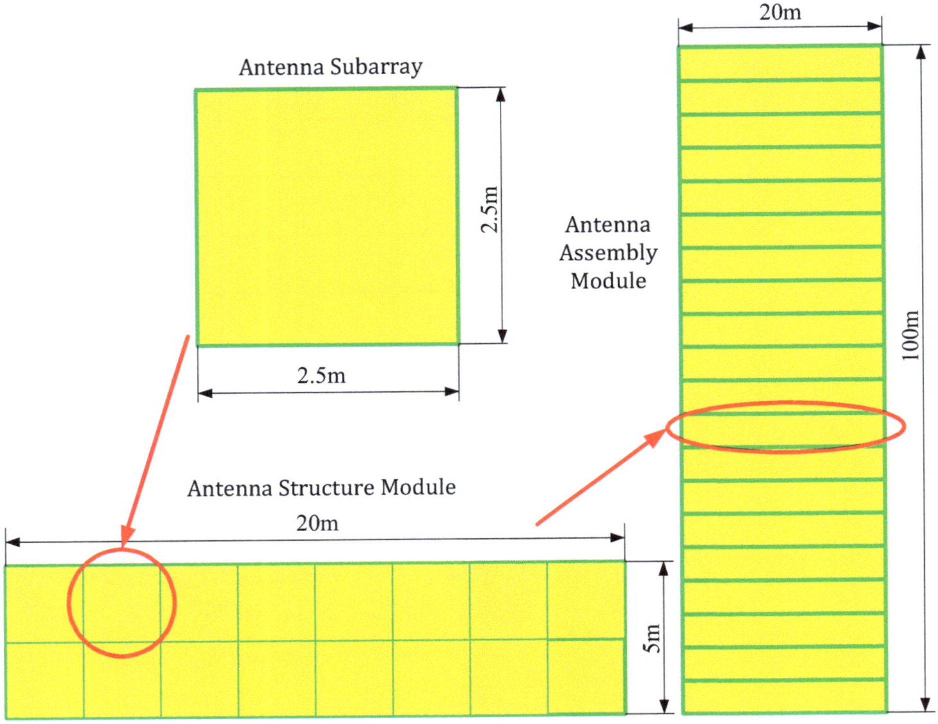

Fig. 7.18 Composition of antenna assembly module

Fig. 7.19 Diagram of antenna subarray division

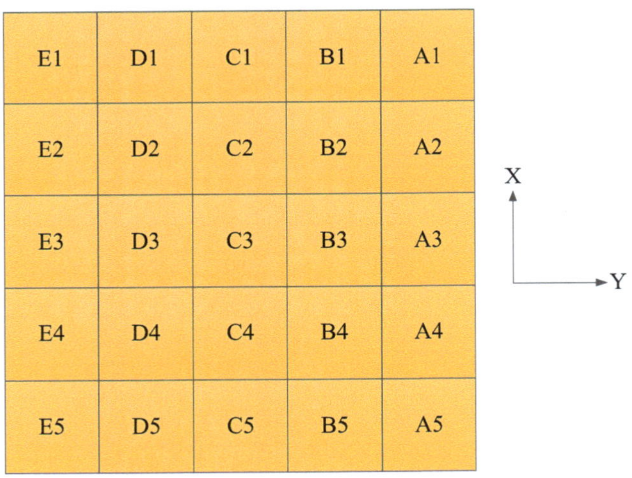

bly module is the basic unit of on-orbit assembly of the antenna, and its folded size is determined according to the launch vehicle. Each antenna assembly module includes 20 antenna structure modules, each with a size of 20 m × 5 m × 0.05 m, which is the basic unit of the antenna structure. 20 antenna structure modules are folded together when launched, with a corresponding size of 20 m × 5 m × 1.1 m. Each antenna structure module includes 16 antenna subarrays (2 × 8 arrangement), so the entire microwave transmitting antenna includes 400 antenna assembly modules, or 8000 antenna structure modules, or 128,000 antenna subarrays.

The antenna subarray is the basic component of the microwave transmitting antenna, with a size of

a. Subarray Module without Receiving Pilot Beam

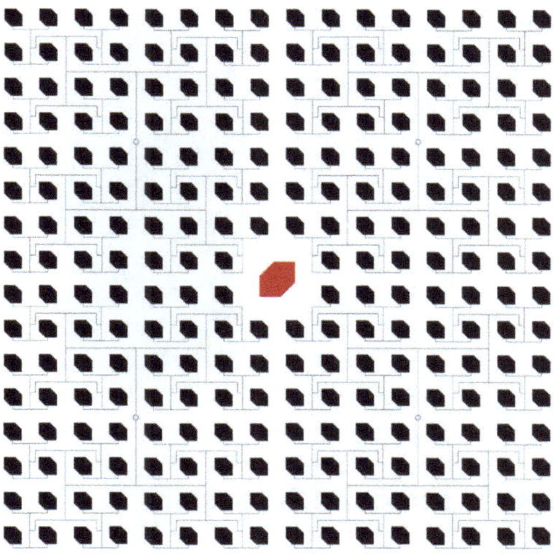

b. Subarray Module Receiving Pilot Beam

Fig. 7.20 Distribution of radiating elements in subarray module. (**a**) Subarray module without receiving pilot beam, (**b**) Subarray module receiving pilot beam

2.5 m × 2.5 m. It is the basic unit of phase control, that is, each antenna subarray corresponds to an independent phase, using a microstrip antenna structure. The antenna subarray uses a fixed feeding network, and the entire subarray is divided into 5 horizontal and 5 vertical total of 25 areas (Fig. 7.19). Each area is a subarray module, with a size of 0.5 m × 0.5 m. The subarray needs to consider the beam pointing requirements in the *Y* direction during design and realize the relative feeding phase difference between areas A and E through the fixed feeding network, thereby achieving the corresponding beam pointing. Each antenna subarray installs a beam direction measurement device which adjusts the phase of each subarray module of the antenna subarray through the beam control unit to compensate for the attitude deviation of the antenna subarray.

The distribution of the radiating elements of the subarray module is shown in Fig. 7.20. Except for C3, each subarray module includes 16 × 16 radiating elements, with a spacing of about 3.125 cm, and all use right-handed circularly polarized microstrip antennas for transmitting microwave. Among them, the center of the C3 subarray module is arranged with a 2.9 GHz left-handed circularly polarized microstrip antenna for receiving the pilot beam to calculate the conjugate phase corresponding to the received signal which will serve as the reference phase of this antenna subarray. The other parts of the radiating elements all use right-handed circularly polarized microstrip antennas, a total of 252, for transmitting microwave beams. Each subarray module contains 2 × 2 microstrip power dividers and each of which contains 8 × 8 basic radiating elements. Each subarray module uses a solid-state power amplifier, which is connected to four microstrip power dividers through a coaxial power divider, and then provides microwave for the 8 × 8 basic radiating elements in the microstrip power divider through the microstrip power divider network.

The power amplifier used for microwave power transmission only needs to cover a single frequency point of 5.8 GHz. The design of the power amplifier can fully utilize its narrowband characteristics to maximize its efficiency and gain. The microwave transmitting antenna includes 128,000 subarrays and each subarray includes 25 subarray modules and requires a total of 3,200,000 power amplifiers. Assuming that the transmitting antenna uses a uniform power intensity distribution and the output power of each power amplifier is 500 W. The power amplifier uses Class F GaN devices with an efficiency of 80%, corresponding to a microwave power of about 1.95 W radiated by each radiating element.

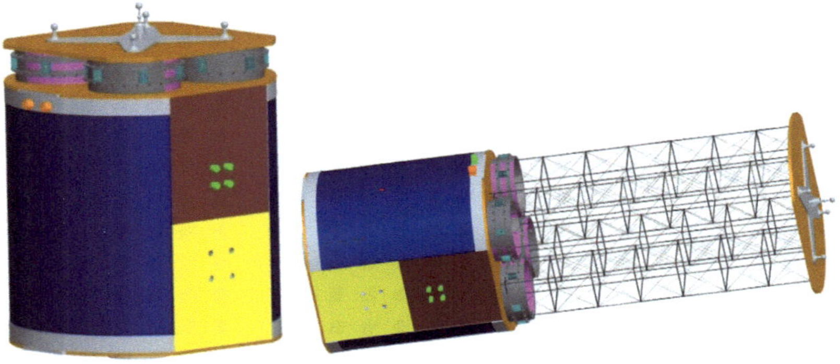

Fig. 7.21 Schematic diagram of main truss module in folded and deployed state

7.4.4 Structure Subsystem

The structural subsystem is composed of various standard structural modules, which form the main structure, solar subarray, and microwave transmitting antenna structure through on-orbit deployment and assembly. The standard structural modules mainly include truss modules, connection modules, and equipment installation platforms.

The main truss module is the basic unit of the main structure and the microwave transmitting antenna structure. It is in a folded state during launch and deployed and locked along the axial direction after being transported into the GEO orbit (Fig. 7.21). The main truss module, according to different assembly requirements, installs docking devices at the front and rear ends or sides for assembly between truss modules or between truss and connection modules. The main truss module is divided into standard type and long type according to the different expansion lengths. The standard main truss module has a diameter of 3 m, a folded length of 3.5 m, a deployed length of about 100 m, and a mass of about 2 t. The long main truss module is mainly used for the hypotenuse part of the microwave antenna array main truss structure and the part connecting the main structure and the microwave antenna array, with a diameter of 3 m, a folded length of 5 m, a deployed length of about 150 m, and a mass of about 3 t. The secondary truss module has the same structure as the main truss module and is mainly used for the support structure of the solar subarray and the microwave transmitting antenna, with a diameter of 2 m, a folded length of 3 m, a deployed length of about 100 m, and a mass of about 1 t.

The connection module can be used for the connection of truss modules at special positions (Fig. 7.22), including: L-shaped connection module, mainly used for the connection of truss modules at the corner of the truss structure, to achieve a 90° angle structure transition; T-shaped connection module, mainly used for the connection of truss modules at the T-shaped position in the truss structure; cross-shaped connection module, used for the connection of truss modules at the cross position in the truss structure; 135° connection module, mainly used for the connection at the corner of the octagonal main truss structure of the microwave antenna array; 5-interface connection module, used for the connection between the main structure truss and the cross-shaped truss of the transmitting antenna array.

The main structure of the SPS is shown in Fig. 7.23. As shown (in the bold blue line section), the main structure connects the solar array and the microwave transmitting antenna together, serving as the main load-bearing structure of the entire SPS together with the solar subarray structure, conductive rotary joint module, and microwave transmitting antenna structure. The main structure mainly includes the north-south main truss structure and the up-down main truss structure, where the north-south main truss structure is used to connect the solar subarray structure, conductive rotary joint, microwave transmitting antenna structure, and the up-down main truss structure is used to connect the two north-south main truss structures.

The solar subarray support structure is a 200 m × 600 m cross structure (the bold yellow line section in Fig. 7.24). The 200 m structure is

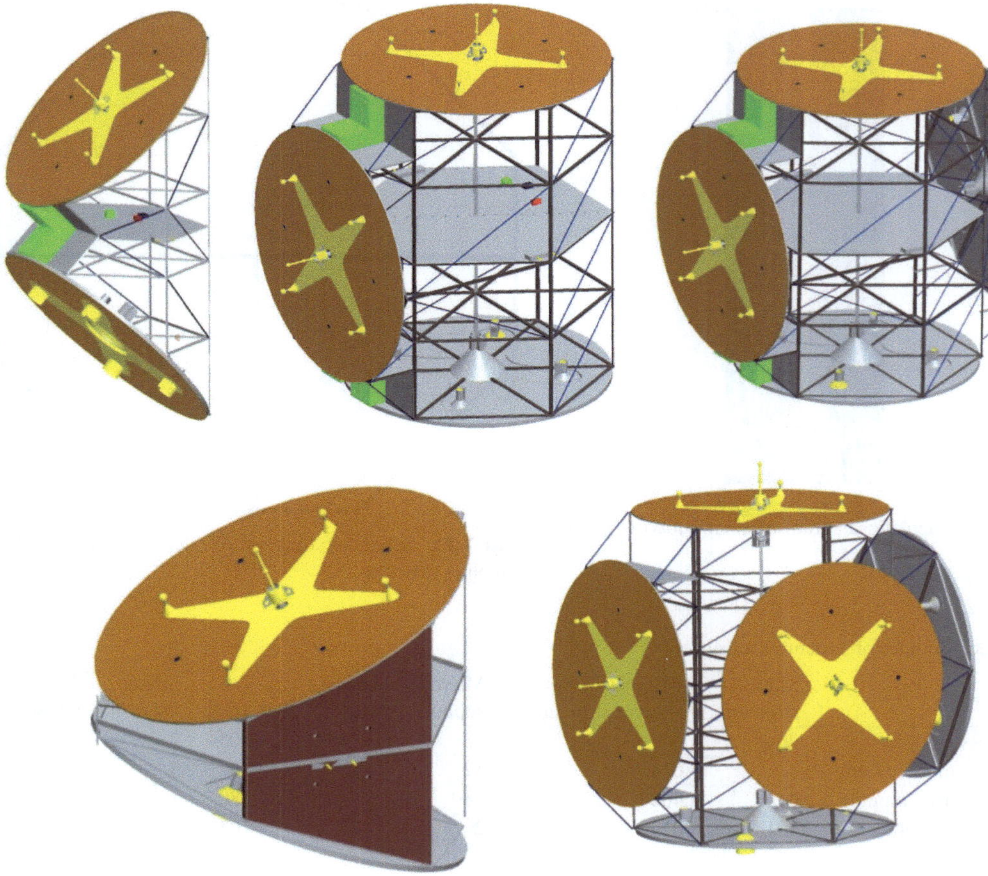

Fig. 7.22 Different connection modules

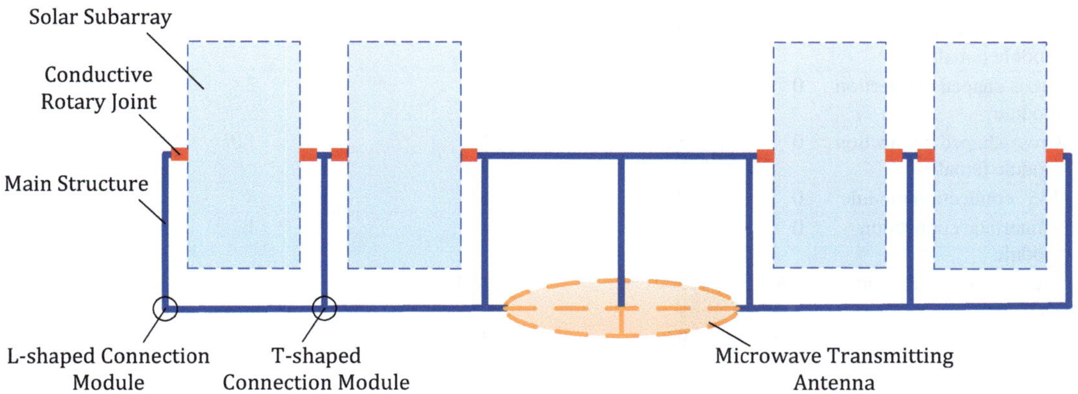

Fig. 7.23 Schematic diagram of main structure

composed of two secondary truss modules, connected to the conductive rotary joints on both sides. The 600 m structure is composed of six secondary truss modules, mainly connected to the solar subarrays, and the middle part is connected by a small cross-shaped connection module.

The microwave transmitting antenna adopts an octagonal configuration (Fig. 7.25), including the

Fig. 7.24 Schematic diagram of solar subarray support structure

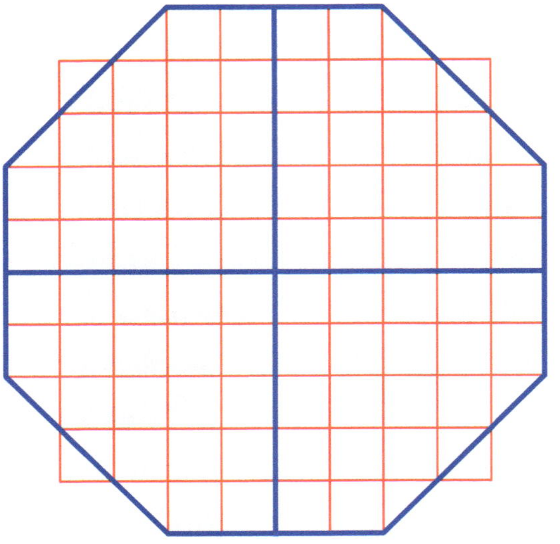

Fig. 7.25 Schematic diagram of microwave transmitting antenna structure

Table 7.3 Modules of structural subsystem (1 GW)

Module	Main structure	Solar subarray support structure	Microwave antenna support structure	Total	Single mass/t	Total mass/t
Main truss module	269	0	36	305	2	610
Main truss module (long)	4	0	12	16	3	48
Secondary truss module	0	400	180	580	1	580
T-shaped connection module	101	0	2	103	0.25	25.75
T-shaped connection module (small)	0	0	20	20	0.12	2.4
L-shaped connection module	4	0	0	4	0.2	0.8
L-shaped connection module (small)	0	0	12	12	0.1	1.2
Cross-shaped connection module	0	0	2	2	0.3	0.6
Cross-shaped connection module (small)	0	50	69	119	0.15	17.85
135° connection module	0	0	8	8	0.2	1.6
5 Interface connection module	0	0	1	1	0.35	0.35
Equipment installation platform	500	500	800	1800	0.1	180
Total	878	950	1142	2970	–	1468.55

main truss structure and secondary truss structure of the microwave transmitting antenna. Considering the connection of the main truss structure and the secondary truss structure, the side lengths of the octagon are 400 m and 424 m, respectively. Among them, the main truss structure is the bold blue line section in the figure, including the octagonal structure and the cross structure, used to connect with the main structure and provide basic mechanical support for the entire antenna. The secondary truss structure is the thin red line section in the figure, the overall structure is a grid format, used to install the microwave transmitting antenna module and related equipment.

The composition of the 1 GW SPS structure subsystem is shown in Table 7.3.

7.4.5 Summary

The preliminary summary of the 1 GW MR-SPS is shown in Table 7.4.

7.5 Transportation of MR-SPS

The total mass of a GW-level MR-SPS is about 10,000 t. The ground-to-LEO transportation sys-

Table 7.4 Specifications of 1GW MR-SPS

System parameters	Operating orbit	GEO
	Assembly orbit	GEO
	Power generation	~1 GW (Maximum)
	System efficiency	~12.6%
	Total mass	~9428 tons
Solar energy collection and conversion subsystem	Solar cells	Thin-film gallium arsenide cells
	Conversion efficiency	~40%
	Solar array area	~6 km^2
	Solar subarray area density	0.3 kg/m^2
	Subarray voltage	~500 V
	Subarray current	~8 kA
	Number of subarrays	600
	Number of arrays	50
	Total weight	~1800 tons
Microwave wireless power transmission subsystem	Microwave frequency	5.8 GHz
	Conversion efficiency	~54%
	Microwave transmitting antenna diameter	1 km
	Antenna type	Phased array antenna
	Number of modules	128,000
	Output power of a module	12.5 kW
	Total mass	3600 tons
	Ground antenna size (diameter)	4.5 km × 6.5 km
Power transmission and management subsystem	Power supply mode	Centralized + distributed
	Main bus voltage for power transmission	20 kV
	Subarray bus voltage	5 kV
	Bus weight	1260 tons
	Microwave power supply voltage	5 kV
	Battery capacity	30 MWh
	Battery weight	60 tons
	Rotary joints weight	100 tons
	Total weight	~2200 tons
Structural subsystem	Supporting structure	Truss deployment
	Total mass	~1468 tons
Attitude and orbit control subsystem	Attitude and orbit control engine	Electric thruster
	Thrust	1.2 N
	Number of engines	340
	Number of momentum wheels	50
	Total mass	60 tons
Others	Thermal control subsystem mass (about 2% of total mass)	200 tons
	Information and system operation management subsystem mass (about 1% of total mass)	100 tons
Ground Receiving Station	Number	1

Table 7.5 Main transportation capabilities of CZ-5 and CZ-9 heavy launch vehicles

Rocket	LEO (0° Inclination)/t	GTO/t	GEO/t	Payload diameter/m	Payload length/m
CZ-5	25	14	5.1	5	20
CZ-9	120	66	–	10	22

Table 7.6 Statistics of the main module status of MR-SPS

Module	Quantity	Folded size/m	Single weight/kg	Total weight/t
Main truss module	305	$\Phi3 \times 3.5$	2000	610
Main truss module (long)	16	$\Phi3 \times 5$	3000	48
Secondary truss module	580	$\Phi2 \times 3$	1000	580
T-shaped connection module	103	$4 \times 3.5 \times 3$	250	25.75
T-shaped connection module (small)	20	$3 \times 2.5 \times 2$	120	2.4
L-shaped connection module	4	$4.7 \times 2.2 \times 3$	200	0.8
L-shaped connection module (small)	12	$3.5 \times 1.5 \times 2$	100	1.2
Cross-shaped connection module	2	$4 \times 4 \times 3$	300	0.6
Small cross-shaped connection module	119	$3.1 \times 3.1 \times 2$	150	17.85
135° connection module	8	$2.8 \times 2.8 \times 3$	200	1.6
5 Interface connection module	1	$4 \times 4 \times 3.5$	350	0.35
Solar subarray module	600	$4 \times 4 \times 2$	3000	1800
Microwave antenna assembly module	400	$20 \times 5 \times 1.1$	9000	3600
Electronic equipment installation platform	1800	–	100	180
Power transmission and management subsystem equipment	–	–	–	2220
Attitude and orbit control subsystem	–	–	–	60
Thermal control subsystem	–	–	–	200
Information and system operation management subsystem	–	–	–	50

tem is considering using CZ-5 launch vehicle and CZ-9 launch vehicle, assuming the main transportation capabilities that can be achieved are shown in Table 7.5. Inter-orbit transportation uses a reusable transporter based on solar electric propulsion.

The main modules of the MR-SPS include the main truss module, secondary truss module, various connection modules, solar subarray module, microwave antenna assembly module, other modules and service system equipment, etc. The corresponding module quantity, size, and weight are shown in Table 7.6. The packaging status of each main module is as follows.

1. Encapsulation state of the main truss modules during launch

The encapsulation state of the main truss modules inside the fairing is shown in Fig. 7.26.

When launched using CZ-5, considering envelope restrictions, one layer can carry 1, a total of 5 layers can be loaded, and 5 main truss modules can be launched at a time, with a total mass of 10–15 t. Assuming the adapter structure mass is 1 t, it meets the launch capability of CZ-5. Approximately 64 launches are required for 321 modules.

If launched using the CZ-9 heavy-lift launch vehicle, according to the envelope limit, one layer can load 7, and a total of 5 layers can be accommodated (one layer is the long type), 35 main truss modules can be launched at a time, with a maximum total mass of 77 t. Assuming the adapter structure mass is 3 t, it meets the launch capability of CZ-9, and 321 modules require approximately 9 launches in total.

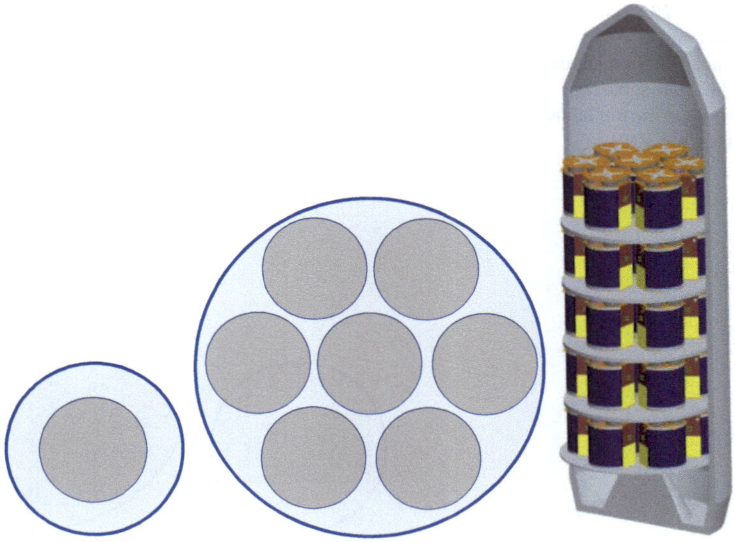

Fig. 7.26 Schematic diagram of main truss module encapsulation state during launch (CZ-5, CZ-9)

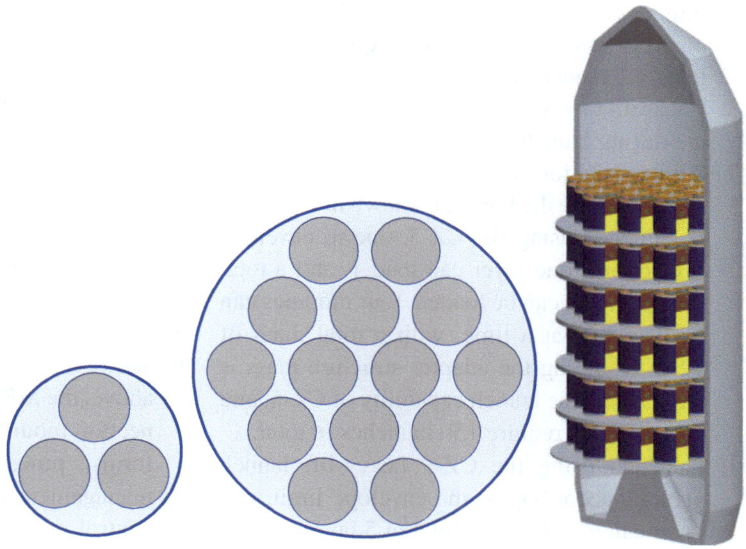

Fig. 7.27 Schematic diagram of encapsulation status of the secondary truss module during launch (CZ-5, CZ-9)

CZ-9)

2. Encapsulation state of the secondary truss module during launch

The encapsulation state of the secondary truss module inside the fairing is shown in Fig. 7.27. If launched using the CZ-5, due to envelope restrictions, one layer can load 3, and a total of 6 layers can be loaded, 18 solar subarray truss modules can be launched at a time, with a total mass of 18 t. Assuming the adapter structure mass is 2 t, it meets the launch capability of CZ-5, and 580 modules require approximately 32 launches in total.

If launched using the CZ-9 heavy-lift launch vehicle, according to the envelope limit, one layer can load 14, and a total of 6 layers can be accommodated, 84 solar subarray truss modules can be launched at a time, with a total mass of 84 t. Assuming the adapter struc-

Fig. 7.28 Schematic diagram of encapsulation status of the solar subarray module during launch (CZ-5, CZ-9)

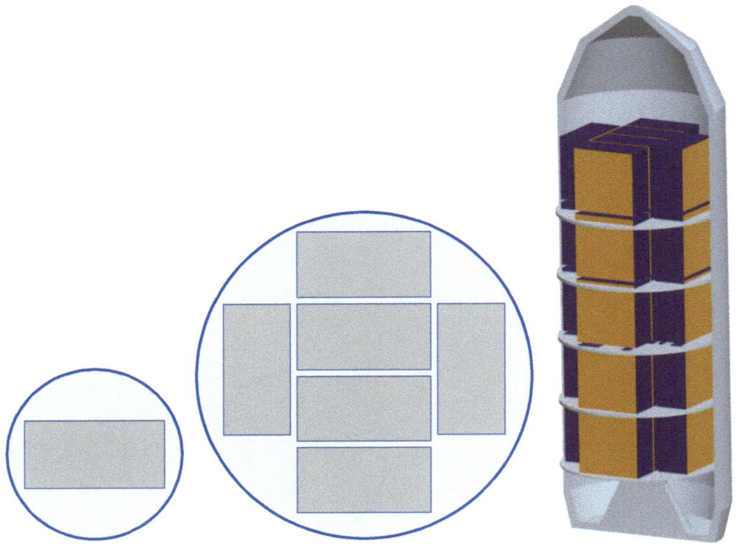

ture mass is 5 t, it meets the launch capability of CZ-9, and 580 modules require approximately 7 launches in total.

3. Encapsulation state of the solar subarray module during launch

The encapsulation state of the solar subarray module inside the fairing is shown in Fig. 7.28. If launched using the CZ-5, due to envelope restrictions, one layer can load 1, and a total of four layers can be loaded, four modules can be launched at a time, with a total mass of 12 t. Assuming the adapter structure mass is 1 t, it meets the launch capability of CZ-5, and 600 modules require 150 launches in total.

If launched using the CZ-9 heavy-lift launch vehicle, according to the envelope limit, one layer can load 6, and a total of 5 layers can be accommodated, 30 modules can be launched at a time, with a total mass of 90 t. Assuming the adapter structure mass is 5 t, it meets the launch capability of CZ-9, and 600 modules require approximately 20 launches in total.

4. Encapsulation state of the microwave antenna assembly module during launch

The microwave antenna assembly module can only be launched using the CZ-9 heavy-lift launch vehicle, and 9 antenna assembly modules can be launched at a time (Fig. 7.29), with a total mass of 81 t. Assuming the adapter structure mass is 5 t, it meets the launch capability, and 400 modules require approximately 45 launches in total.

5. Others

In addition to the main components mentioned above, the MR-SPS also includes various connection modules, equipment installation platforms, power transmission cables, power management equipment, electric thrusters, control system-related equipment, and information and system operation management equipment, with a total mass of about 2760 t. If launched using the CZ-5 rocket, each launch transports an average of 20 t, requiring approximately 138 launches in total. If

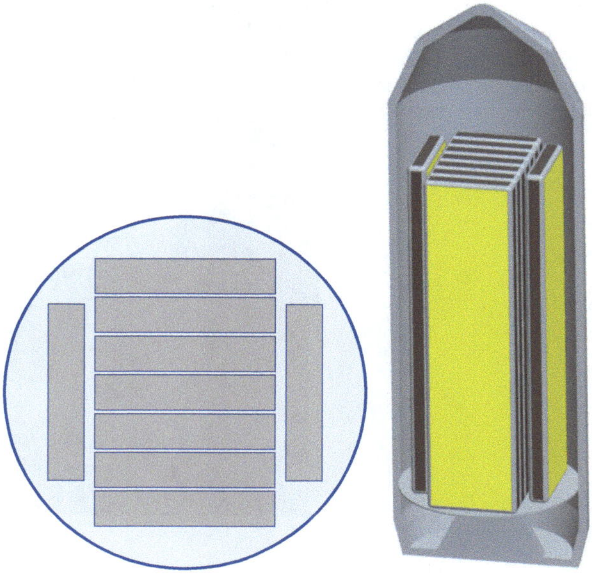

Fig. 7.29 Schematic diagram of encapsulation status of the microwave antenna assembly module during launch (CZ-9)

Table 7.7 Statistics of the number of transport times for the main modules of an MR-SPS

	Main truss module	Secondary truss module	Solar subarray module	Microwave antenna assembly module	Other
Quantity by CZ-5 per time	5	18	4	–	–
Launch number of CZ-5	64	32	150	–	138
Quantity by CZ-9 per time	35	84	30	9	–
Launch number of CZ-5	9	7	20	45	28
Total launch number of CZ-5	384 (excluding antenna launches)				
Total launch number of CZ-9	109				

launched using the CZ-9 heavy-lift launch vehicle, each launch transports an average of 100 t, requiring approximately 28 launches in total.

The statistics of the number of transport times for the main modules of an MR-SPS are shown in Table 7.7.

After the assembly modules of the SPS are launched into LEO by heavy-lift launch vehicles, they are transported to GEO by a reusable solar electric propulsion inter-orbit transporter. The inter-orbit transporter first operates in LEO, and after the assembly module separates from the launch vehicle, it is grabbed by the mechanical arm of the inter-orbit transporter and docked (Fig. 7.30). Then the inter-orbit transporter's protective cover closes, and the electric propulsion system is started to transport the payload to GEO, where the space assembly modules are released and deployed as needed. At last, it returns to LEO for the next transport according to mission planning.

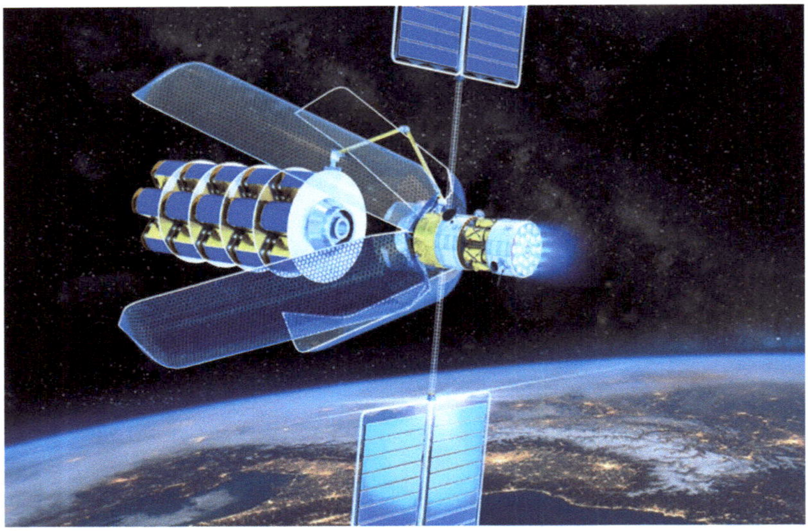

Fig. 7.30 Schematic diagram of the docking payload of the reusable solar electric propulsion inter-orbit transporter

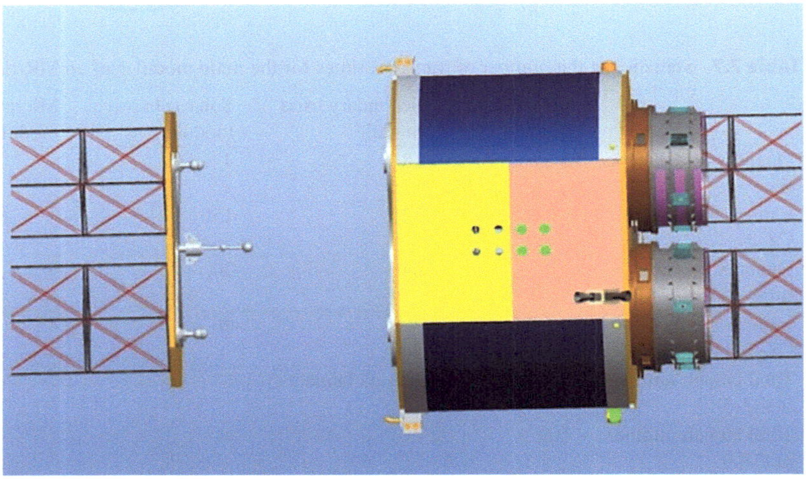

Fig. 7.31 Assembly status of the truss module

7.6 On-Orbit Assembly of MR-SPS

According to the configuration and module design of MR-SPS, the modules that need to be assembled mainly include the main truss module, the secondary truss module, various truss connection modules, solar subarray modules, microwave antenna assembly modules, the installation of service system equipment and connection cables, etc. The typical assembly status of the modules is as follows.

1. Assembly between truss modules

The assembly between truss modules adopts a docking-first-then-deployment method. Each truss module to be assembled needs to dock with an already deployed truss module or connection module through an assembly robot, and then use its own deployment mechanism for truss deployment (Fig. 7.31).

2. Assembly of connection modules with truss modules

The assembly of connection modules with truss modules is directly docked with the already deployed truss module through the assembly robot. Typical assembly statuses include the assembly status of the L-shaped truss connec-

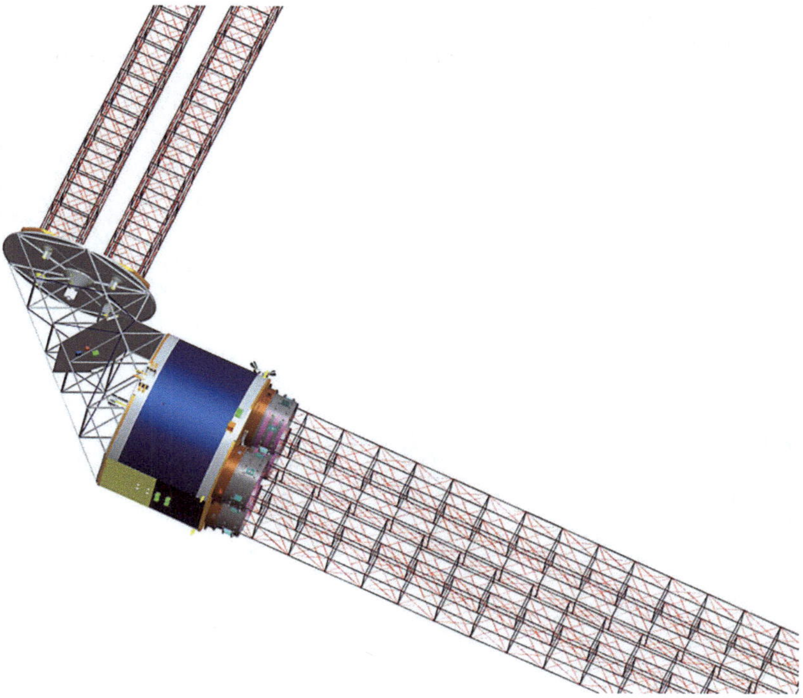

Fig. 7.32 Assembly status of the L-shaped truss connection module

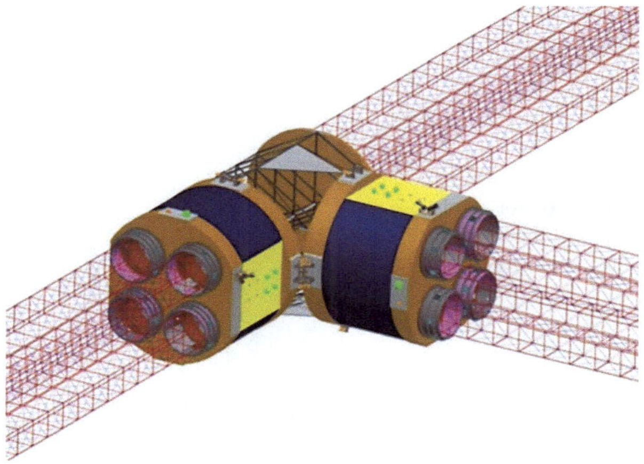

Fig. 7.33 Assembly status of the T-shaped docking module

tion module (Fig. 7.32), T-shaped docking module assembly status (Fig. 7.33), 135° docking module assembly status (Fig. 7.34), cross-shaped truss connection module assembly status (Fig. 7.35), and 5-interface truss connection module assembly status (Fig. 7.36).

3. Assembly between solar subarray modules and secondary truss modules

The assembly between solar subarray modules and secondary truss modules adopts a docking-first-then-deployment method. Each solar subarray module to be assembled needs to dock with a docking interface of the secondary truss module through an assembly robot, and then use its own deployment mechanism and the assistance of the assembly robot to deploy the solar subarray. After deployment, it docks with other modules

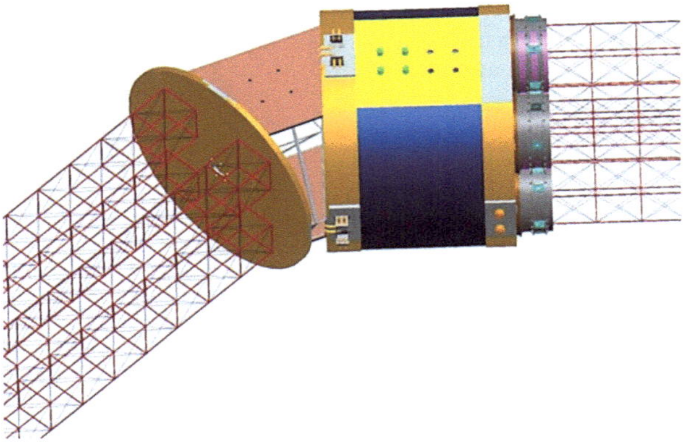

Fig. 7.34 Assembly status of the 135° docking module

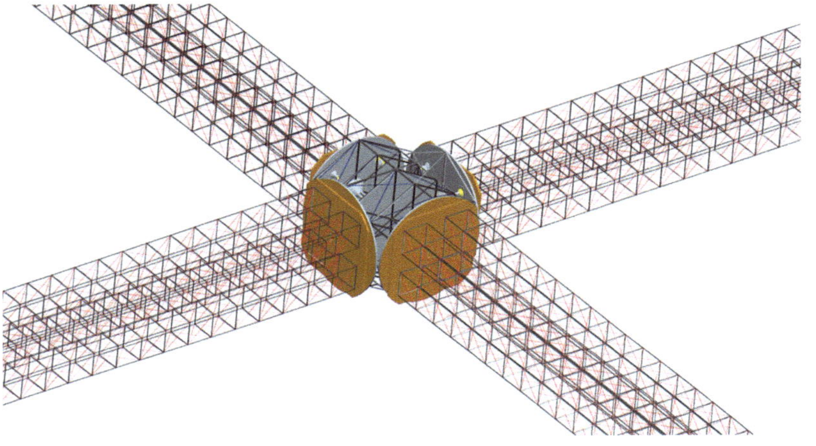

Fig. 7.35 Assembly status of the cross-shaped truss connection module

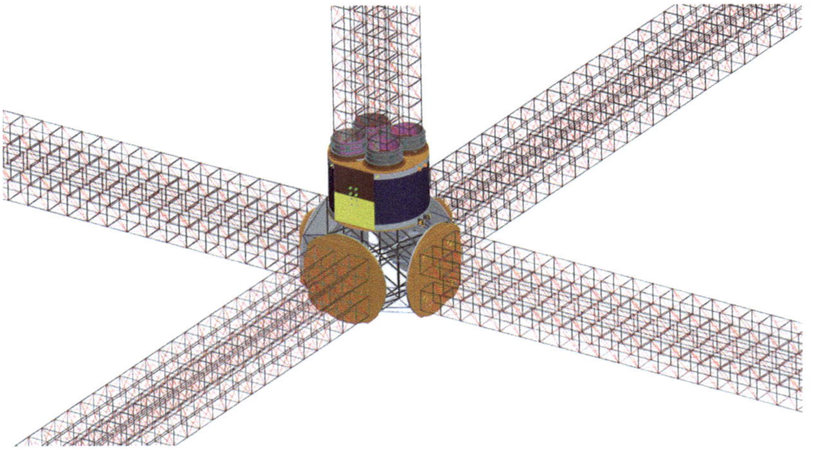

Fig. 7.36 Assembly status of the 5 interface truss connection module

(including truss modules and other solar subarrays). Figure. 7.37 shows the assembly status between the solar subarray module and the secondary truss module.

4. Assembly between microwave antenna assembly modules and secondary truss modules

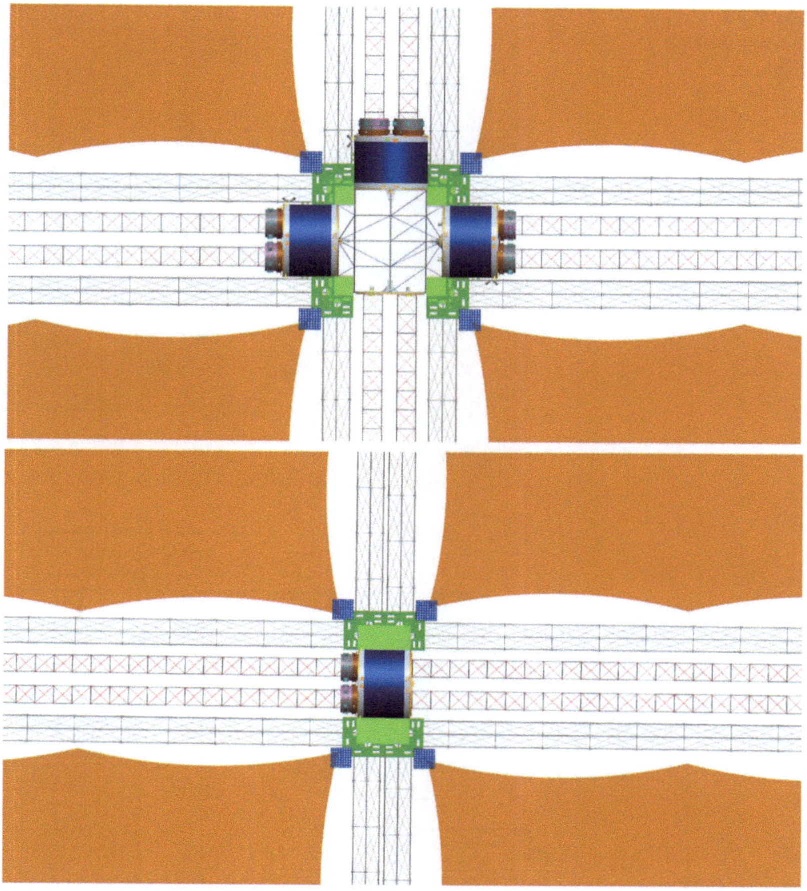

Fig. 7.37 Assembly status between the solar subarray module and the secondary truss module

The microwave antenna assembly module has a heavy mass and large deployment size. The assembly between the microwave antenna assembly module and the secondary truss module adopts a deployment-first-then-docking method. Each microwave antenna assembly module needs to be deployed in orbit through the cooperation of multiple assembly robots, then transported to the designated installation position, and then assembled with the assembly interface of the secondary truss module through the cooperation of multiple assembly robots (Fig. 7.38).

7.7 Economic Aspects of the SSP System

7.7.1 Full Cycle Cost Analysis Method

The economic analysis of SSP considers the direct costs of the entire process from design to end of life, mainly including system design, development and construction, launch deployment, assembly testing, operation and maintenance, and system shutdown and reuse six stages. But it does not include the costs incurred

Fig. 7.38 Laying of antenna assembly modules

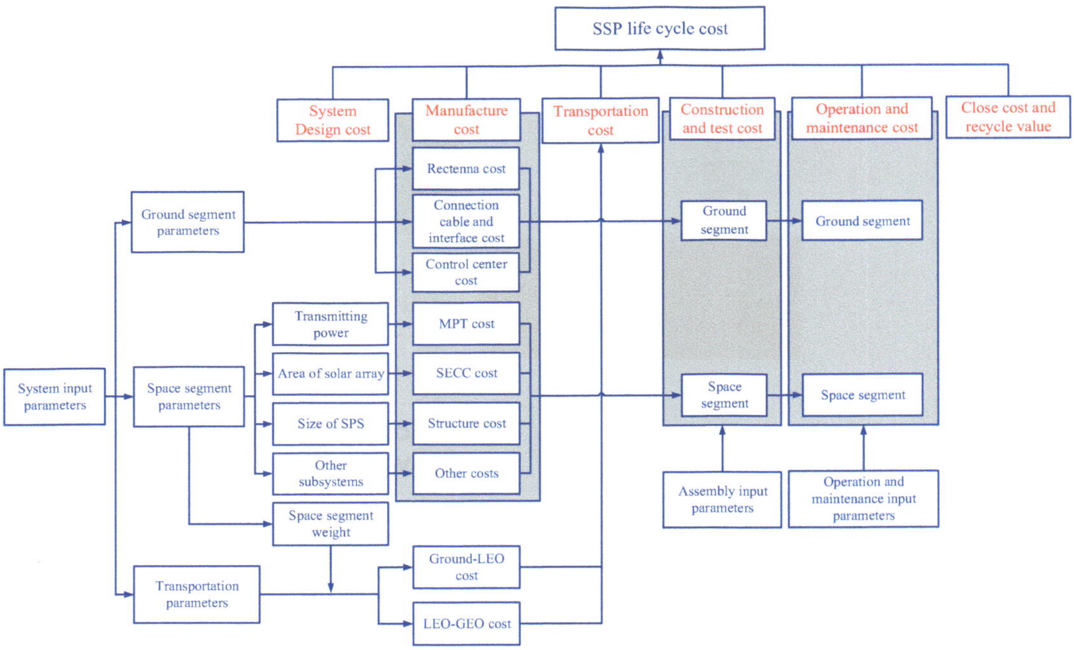

Fig. 7.39 SSP cost analysis method

for the preliminary technical research and development and system demonstration for the development of the SSP system, nor does it include the research and development and infrastructure costs of large systems such as transportation, launch sites, space construction and support, and ground operation control. The cost analysis process of SSP is shown in Fig. 7.39. First, the top-level input parameters of the system need to be clarified, then the relevant parameters of the ground segment and space segment are determined according to the design scheme, and the development costs of the main subsystems are determined based on the relevant parameters. The launch costs of the main subsystems are determined according to the transportation input parameters, and the assembly costs of the SPS are determined according to the assembly input parameters, and then the construction costs of the SPS are determined. The operation and maintenance processing costs of the entire system are determined according to the complexity of on-orbit operation, maintenance, and decommission. The sum of these costs is the full cycle cost of SSP, combined with the total power generation during the life cycle, the power generation cost of SSP can be obtained.

7.7.2 Power Station Cost Analysis Results

The top-level input parameters for the SSP cost analysis are as follows:

1. Power generation capacity: 1 GW (average power)
2. System efficiency: about 13%
3. System operating life: 30 years
4. Launch capability:
   - Ground-LEO (CZ-9 heavy-lift): 120 t (maximum)
   - LEO-GEO: 120 t (maximum)
5. Transportation cost:
   - Ground-LEO: 14 million $/time
   - LEO-GEO: 28 million $/time (including the depreciation cost of the inter-orbit transporter)
6. Launch and assembly cycle: 2 years
7. Size of the receiving antenna: average 5.5 km

Based on the above top-level parameters and the parameters of each subsystem of the SSP system, a preliminary analysis shows that the total cost of research and development, launch, construction, operation, etc. of the SSP system during the entire life cycle is about 18.5 billion US dollars. At the same time, under normal operation, the power generation of a 1 GW SSP system is about 236.5 billion kWh during its 30-year life. This means that the cost of each kWh electricity is $0.078 which is comparable to traditional solar power on ground.

It cannot be denied that the commercial feasibility of space solar energy is currently one of the biggest challenges for its global application. Although the basic physical principles of space solar power were confirmed in the 1960s, the satellite launch technology back then and the huge cost of launching heavy solar arrays into orbit, as well as the low prices of other energy sources, made space solar power impractical in the following decades.

However, in the past decade, with the increasing maturity of recyclable launch vehicle technology and the significant reduction in the weight of solar arrays, the cost of space solar energy has dropped to one tenth or even less of its original value. According to the progress of some countries, in the next 20 years, although space solar power may still be costlier than traditional solar and wind power, it can defeat energy sources that must be filled when renewable energy cannot generate electricity.

Collecting sunlight in space can also help reduce the huge demand for land from traditional solar and wind power generation. They can directly transmit electricity to where it is needed, providing power for remote areas that are difficult to connect to the grid. This capability may also potentially reduce the demand for transmission lines, as the installation cost of transmission lines is high, requires a large amount of raw materials, and may trigger wildfires.

Currently, over 10 countries are accelerating the development of space solar power, including the USA, the UK, Australia, Canada, China, ESA, Japan, Korea, Poland, Saudi Arabia, Switzerland, and UAE. The countries are developing, testing, and planning various types of space solar power satellite systems. The next few years will be a testing ground for these different competitive systems. In the foreseeable future, we can expect SPP to become an important infrastructure providing massive, continuous, and stable green electricity for the entire humanity, not only in developed regions, but of particularly benefit for people in remote areas.

Bibliography

1. Hou X B, Zhang X H, Wang L. Introduction on space solar power[M] (in Chinese). Beijing: China Astronautic Publishing House, 2020

MIX
Papier aus verantwortungsvollen Quellen
Paper from responsible sources
FSC® C105338

If you have any concerns about our products,
you can contact us on
ProductSafety@springernature.com

In case Publisher is established outside the EU,
the EU authorized representative is:
Springer Nature Customer Service Center GmbH
Europaplatz 3, 69115 Heidelberg, Germany

Printed by Libri Plureos GmbH
in Hamburg, Germany